# Communications in Asteroseismology

Volume 160
February, 2010

T0133219

Compendium
March to October, 2009

Austrian Academy
of Sciences Press

Vienna 2010

OAW

# Communications in Asteroseismology

Editor-in-Chief: **Michel Breger**, michel.breger@univie.ac.at
Editorial Assistant: **Isolde Müller**, isolde.mueller@univie.ac.at
Layout & Production Manager: **Paul Beck**, paul.beck@ster.kuleuven.be
Language Editor: **Natalie Sas**, natalie.sas@ster.kuleuven.be

CoAst Editorial and Production Office
Türkenschanzstraße 17, A - 1180 Wien, Austria
*http://www.oeaw.ac.at/CoAst/*
*Comm.Astro@univie.ac.at*

## Cover Illustration

Simulation of A star surface convections
(Illustration kindly provided by F. Kupka and J. Ballot. For more information
see the paper by F. Kupka et al., page 30)

British Library Cataloguing in Publication data.
A Catalogue record for this book is available from the British Library.

Austrian Academy of Sciences Press
A-1011 Wien, Postfach 471, Postgasse 7/4
Tel. +43-1-515 81/DW 3402-3406, +43-1-512 9050
Fax +43-1-515 81/DW 3400
http://verlag.oeaw.ac.at, e-mail: verlag@oeaw.ac.at

*Comm. in Asteroseismology*
*Volume 160, Compendium 2009*
© *Austrian Academy of Sciences*

# Introductory Remarks

Starting with this regular issue (CoAst 160), we are changing our procedure in order to achieve immediate publishing. As soon as a paper is accepted by the referee and the editorial office, it is published electronically through ADS and can also be downloaded from our web site. This also means that the author will be able to cite his/her paper very quickly (and hopefully obtain a raise in salary as in some countries the number of publications count.) All the new accepted and electronically published papers will be collected periodically and distributed in a printed volume. However, the date of publication is the date of the electronic publication, not the printing.

We hope that this will lead to much faster publication. There will no longer be a deadline for our papers; just submit your paper when it is finished. By the way, we are not the only journal changing to this system.

Michel Breger
Editor-in-Chief

# Contents

# Scientific
# Papers

Comm. in Asteroseismology
Volume 160, March 2009
© Austrian Academy of Sciences

# Observations of candidate oscillating eclipsing binaries and two newly discovered pulsating variables

A. Liakos, and P. Niarchos

National and Kapodistrian University of Athens, Faculty of Physics,
Department of Astrophysics, Astronomy and Mechanics,
GR 157 84, Zografos, Athens, Greece

## Abstract

CCD observations of 24 eclipsing binary systems with spectral types ranging between A0-F0, candidate for containing pulsating components, were obtained. Appropriate exposure times in one or more photometric filters were used so that short-periodic pulsations could be detected. Their light curves were analyzed using the Period04 software in order to search for pulsational behaviour. Two new variable stars, namely GSC 2673-1583 and GSC 3641-0359, were discovered as by-product during the observations of eclipsing variables. The Fourier analysis of the observations of each star, the dominant pulsation frequencies and the derived frequency spectra are also presented.

Accepted:    2009, March 16
Individual Objects:   AD And, RY Aqr, UW Boo, AL Cam, AY Cam, TY Cap, EG Cep, RW Cet, WY Cet, BR Cyg, KR Cyg, V456 Cyg, V466 Cyg, V1034 Cyg, RZ Dra, Z Dra, TZ Eri, GSC 2673-1583, GSC 3641-0359, TX Her, V338 Her, V359 Her, IQ Per, GH Peg, X Tri, RS Tri

## 1.   Introduction and Observations

The subject of the present study is a search for $\delta$ *Scuti* components in several detached and semi-detached systems. Most of these components were selected from the list of Soydugan et al. (2006). Mkrtichian et al. (2004) introduced the oEA (oscillating EA) stars as the (B)A-F spectral type mass-accreting main-sequence pulsating stars in semi-detached Algol-type eclipsing binary systems.

In general, in order to search photometrically for possible short periodic pulsations in a single star or a binary system, the following observational guidelines should be followed: i) the span time should be longer than 2.5 hrs, since the majority of the pulsation periods of $\delta$ Scuti stars range between 20 min-8 hrs (Soydugan et al. 2006), ii) the B-filter should be used since the pulsations have their highest amplitude in this spectral range, given that the pulsating components are of early spectral type (A2-F2). For the brighter systems, other filters could also be used. iii) the comparison star should be bright enough in order to decrease the standard deviation of each observation, iv) appropriate exposure times must be used in order to achieve a photometric SNR (signal-to-noise ratio) higher than 100.

In addition to the general observational guidelines mentioned above, the relatively small diameter (0.4m) of the telescope used for the observations of the present work, puts an upper limit of 11 mag for the observations in the B-filter.

The observations were carried out during June 2007-December 2008 at the Gerostathopoulion Observatory of the University of Athens, using the 40cm Cassegrain telescope equipped with the ST-8XMEI CCD camera and the BVRI Bessell photometric filters.

In some cases (AD And, AL Cam, X Tri), although the results for possible pulsational behaviour were negative (see table 1), the observations were continued in order to obtain complete light curves which will be used for future work, unrelated to the present study.

The star GSC 2673-1583 ($\alpha_{2000} = 19^h54^m39.5^s$, $\delta_{2000} = +32°56'2.7''$, $V_{mag} = 10.8$) was detected as a variable by Liakos & Niarchos (2008) in the field of view of V466 Cyg and it was observed during August and September of 2008 for 6 nights using the B-filter. The star GSC 3641-0359 ($\alpha_{2000} = 23^h35^m50^s, \delta_{2000} = +48°43'43'', V_{mag} = 11.3$) was observed using BVRI filters for 5 nights during the same time period, while its variability was also discovered by the authors in the field of view of AD And. The finding charts of both stars are presented in figure 2.

## 2. Results of frequency analyses

Differential magnitudes were obtained for all systems using the software Muniwin v.1.1.23 (Hroch 1998). We performed frequency-analysis of all the available observational points in the interval from 5 to 80 c/d (typical range for $\delta$ Scuti stars, Breger 2000) with the software PERIOD04 which is based on the classical Fourier analysis (Lenz & Breger 2005). In table 1 we list the name of the system given in the GCVS catalog, the apparent magnitude and the spectral type of the system, the filters used, the number of nights of observations, the total

Table 1: The log of observations and the results for the possible pulsational behaviour of the selected 24 eclipsing binaries

| System | appar. mag | Spec. type | filters | nights | time span (hrs) | SD (mmag) | phase interval | $f_{dom}$ | ref. |
|---|---|---|---|---|---|---|---|---|---|
| AD And | 11.2 (V) | A0 | BVRI | 5 | 24 | 4.3 | 0.0 1.0 | – | 2 |
| RY Aqr | 8.9 (V) | A8 | B | 1 | 3 | 5.4 | 0.18-0.25 | – | 1 |
| UW Boo | 10.4 (B) | F0 | B | 1 | 5 | 5.7 | 0.43-0.63 | – | 1 |
| AL Cam | 10.3 (V) | A2 | BV | 7 | 34 | 6.8 | 0.0-1.0 | – | 1 |
| AY Cam | 9.9 (B) | – | BV | 1 | 6 | 4.9 | 0.73-0.82 | – | – |
| TY Cap | 10.3 (V) | A5 | BVRI | 12 | 37 | 5.2 | 0.0-1.0 | 24.2 c/d | 1 |
| EG Cep | 9.5 (V) | A3 | BVR | 2 | 13 | 4.1 | 0.0-1.0 | – | 1 |
| RW Cet | 10.1 (B) | A5 | B | 1 | 5.5 | 3.8 | 0.69-0.92 | – | 1 |
| WY Cet | 9.6 (B) | A2 | BVRI | 16 | 52 | 2.8 | 0.0-1.0 | 13.2 c/d | 2 |
| BR Cyg | 9.9 (V ) | A5 | B | 1 | 4 | 3.9 | 0.79-0.91 | – | 1 |
| KR Cyg | 9.3 (V) | A1 | B | 2 | 10.6 | 3.9 | 0.18-0.36, 0.42-0.75 | – | 2 |
| V456 Cyg | 10.8 (V) | A2 | B | 1 | 7.5 | 4.2 | 0.17-0.51 | – | 1 |
| V466 Cyg | 10.5 (V) | A8 | B | 6 | 27.5 | 3.1 | 0.15-0.5, 0.55-0.82 | – | 1 |
| V1034 Cyg | 9.6 (B) | A0 | B | 1 | 3.2 | 3.2 | 0.63-0.76 | – | 2 |
| RZ Dra | 10.4 (V) | A5 | B | 1 | 3.2 | 4.1 | 0.62-0.86 | – | 1 |
| Z Dra | 10.6 (V) | A5 | V | 1 | 3 | 4.1 | 0.68-0.77 | – | 1 |
| TZ Eri | 9.8 (V) | A5/6 | BV | 26 | 70 | 2.7 | 0.0-1.0 | 18.7 c/d | 1 |
| TX Her | 8.1 (V) | A8 | B | 1 | 3 | 6.5 | 0.79-0.86 | ? | 1 |
| V338 Her | 10.2 (V) | A9 | B | 3 | 19.4 | 4.1 | 0.16-0.60, 0.63-0.81 | ? | 1 |
| V359 Her | 10.0 (V) | F0 | BV | 2 | 10 | 7.1 | 0.16-0.32, 0.58-0.65 | – | 1 |
| IQ Per | 7.73 (V) | A7 | BV | 1 | 4.5 | 4.4 | 0.10-0.21 | – | 1 |
| GH Peg | 9.1 (B) | A3 | BVR | 2 | 10.3 | 3.9 | 0.22-0.28, 0.63-0.74 | – | 1 |
| X Tri | 9.0 (V) | A3 | BVR | 13 | 25.3 | 2.8 | 0.0-1.0 | – | 1 |
| RS Tri | 10.3 (V) | A5 | B | 2 | 8 | 4.3 | 0.12-0.18, 0.66-0.78 | – | 1 |

1-Soydugan et al. (2006), 2-Kreiner et al. (2001)

time span, the standard deviation (SD) of the observations (mean value), the phase intervals, the dominant pulsating frequency found and the reference used for quoting the spectral types and the apparent magnitudes of each system. For the majority of the systems, the observations were made in the phase interval where the components were at maximum separation, namely around maxima. Moreover, for the systems which seem to have a pulsating component, we list the most dominant pulsation frequency, while for the ambiguous ones, marked by a (?), the results of frequency analysis are not very convincing, since the possible pulsation frequencies were found to have a SNR<4.0, which is less than the pulsation detection limit (Lenz & Breger 2005) and an amplitude inside the SD limit. The light curves of the largest (data) sets of observations for certain cases of oEA stars are illustrated in figure 1.

## 3.   The two newly discovered variables

The data reduction and the frequency analysis methods for studying the new variables are the same as for the eclipsing binaries mentioned above. For GSC 3641-0359 we used the stars GSC 3641-0045 as comparison star (C) and GSC 3641-0037 as check star (K), while for GSC 2673-1583, the GSC 2673-3481 and

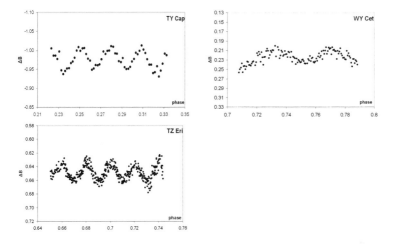

Figure 1:  The partial light curves of the systems found to have a pulsating component.

GSC 2673-2979 were used as (C) and (K) stars, respectively.  After the first
frequency computation, the residuals were prewhitened for the next one.  The
results of the frequency search for both stars are listed in table 2, listing the
identification number of the frequency, the frequency value with its error and
the signal-to-noise ratio (SNR) after prewhitening for the previous frequency/-
ies.  The corresponding amplitudes and their standard deviations, derived from
a multi-parameter least-squares fit of sinusoidal functions, are also given.  The
frequency spectra and the Fourier fits to the observational points for the largest
(data) sets of observations of each star are presented in figures 3-4.

## 4.   Discussion and conclusions

The aim of the present study is to present the results for the pulsational behaviour of some selected eclipsing binaries and for the discovery of two new
variable stars.  CCD observations of 24 systems, candidate for including pulsating components, have been obtained and analysed using Fourier techniques to
search for short-periodic pulsations.  The analysis was performed on all available
observations.  For the brighter targets, we have observations in more than one
photometric filter.  Nineteen of them do not show any clear evidence of high-
frequency pulsations.  V338 Her and TX Her are not very convincing cases as
the analysis revealed frequencies with amplitudes inside the error limit (SD) and
a signal-to-noise ratio less than 4.5.  Finally, the systems TY Cap, WY Cet and
TZ Eri are clear cases of oEA stars and their dominant frequencies are given in

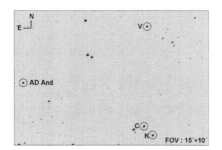 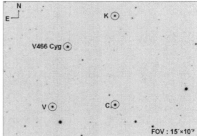

Figure 2: The finding charts of the stars GSC 3641-0359 (V in the left part) and GSC 2673-1583 (V in the right part). The dimensions and the orientation of the field of view and the comparison (C) and check (K) stars are also indicated.

Table 2: The results of frequency analysis for the new variables

| | | GSC 3641-0359 | | | GSC 2673-1583 | | |
|---|---|---|---|---|---|---|---|
| Filter | No | Freq. (c/d) | Ampl. (mmag) | SNR | Freq. (c/d) | Ampl. (mmag) | SNR |
| B | $f_1$ | 8.3143(4) | 15.1(5) | 13.56 | 10.0280(4) | 9.9(2) | 16.20 |
| | $f_2$ | 11.2825(7) | 8.0(5) | 6.32 | 0.4906(5) | 7.4(2) | 8.34 |
| | $f_3$ | 7.6710(10) | 5.7(5) | 5.36 | 8.8572(12) | 3.1(2) | 4.94 |
| | $f_4$ | 1.2785(17) | 3.3(5) | 6.30 | | | |
| V | $f_1$ | 8.3126(6) | 10.3(5) | 11.65 | | | |
| | $f_2$ | 11.7081(11) | 5.6(5) | 5.73 | | | |
| | $f_3$ | 7.6700(13) | 4.6(5) | 5.25 | | | |
| | $f_4$ | 1.0119(14) | 4.2(5) | 6.59 | | | |
| R | $f_1$ | 8.8859(6) | 9.6(5) | 13.07 | | | |
| | $f_2$ | 11.1346(12) | 4.6(5) | 5.25 | | | |
| | $f_3$ | 7.5222(20) | 2.9(5) | 4.08 | | | |
| | $f_4$ | 0.9887(72) | 7.9(5) | 10.11 | | | |
| I | $f_1$ | 8.8858(5) | 9.1(4) | 20.26 | | | |
| | $f_2$ | 11.0324(9) | 5.0(4) | 13.65 | | | |
| | $f_3$ | 7.9523(13) | 3.7(4) | 6.88 | | | |
| | $f_4$ | 0.8454(9) | 5.2(4) | 10.80 | | | |

table 1. The pulsational behaviour of TZ Eri has been detected by Mkrtichian et al. (2006) and a frequency analysis in detail has been published by Liakos et al. (2009). The detailed analysis of TY Cap and WY Cet will be presented in future work. All the systems were observed near their maxima phases, for at least 3 hours and in some cases, the whole binary's light curve was obtained.

Further observations, using larger telescopes, are certainly needed to clarify the pulsational behaviour of the non-convincing cases.

During the observations of V466 Cyg and AD And, two new variables have been detected and their observations were analysed using the same method. The Fourier analysis revealed that both cases show pulsational behaviour on

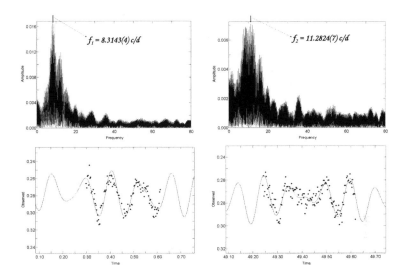

Figure 3: The frequency spectra (upper panels) and the Fourier fits (lower panels) to the observational points for the longest (data) sets of observations of GSC 3641-0359.

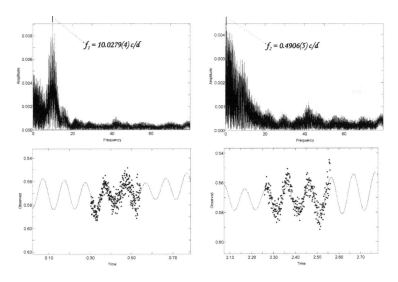

Figure 4: The same as figure 3 for GSC 2673-1583.

multiperiodic mode. Due to the lack of spectroscopic information, they cannot be categorized. Their pulsational characteristics are similar to those of $\delta$ *Scuti* stars (5-80 c/d, Breger 2000) and also to those of $\beta$ *Cephei* stars. We did not find any reasonable frequency values with amplitudes smaller than 2.9 mmag and a signal-to-noise ratio higher than 4, after prewhitening for the last frequencies in each filter.

GSC 3641-0359 was found to have four pulsation frequencies with the dominant one 8.3143(4) c/d (with an amplitude of 15.1(5) mmag), while the period analysis of GSC 2673-1583 revealed a tri-periodic behaviour dominated by the frequency 10.0280(4) c/d (with an amplitude of 9.9(2) mmag). However, the frequencies $f_4$ of GSC 3641-0359 and $f_2$ of GSC 2673-1583, since they have a value near 1 c/d, seem to correspond to g-mode pulsations, and could be caused by rotational effects of the star or even by systematic observational errors relevant to atmospheric extinctions (Breger 2005). Of course, the tidal forces of a close companion could also cause these g-mode frequencies, but so far there is no evidence that these stars are members of binary systems. Further spectroscopic observations will certainly reveal the spectral type of each star and will help to detect any close companion(s).

**Acknowledgments.** We thank the anonymous referee for valuable comments and suggestions that improved the paper. This work has been financially supported by the Special Account for Research Grants No 70/4/5806 of the National & Kapodistrian University of Athens, Greece.

## References

Breger, M. 2000, ASPC, 210, 3

Breger, M. 2005, ASPC, 333, 138

Hroch, F. 1998, Proceedings of the 29th Conference on Variable Star Research, 30

Kreiner, J. M., Kim, C.-H., & Nha, I.-S. 2001, An Atlas of O-C Diagrams of Eclipsing Binary Stars / by Jerzy M. Kreiner, Chun-Hwey Kim, Il-Seong Nha. Cracow, Poland: Wydawnictwo Naukowe Akademii Pedagogicznej. 2001.

Mkrtichian, D. E., Kusakin, A. V., Rodriguez, E., et al. 2004, A&A, 419, 1015

Mkrtichian, D., Kim, S.-L., Kusakin, A. V., et al. 2006, ApSS, 304, 169

Lenz, P., & Breger, M. 2005, CoAst, 146, 53

Liakos, A., & Niarchos, P. 2008, IBVS, 5900, in press

Liakos, A., Ulaş, B., Gazeas, K., & Niarchos, P. 2009, CoAst, 157, 336

Soydugan, E., Soydugan, F., Demircan, O., & İbanoğlu, C. 2006, MNRAS, 370, 2013

Comm. in Asteroseismology
Volume 160, August 2009
© Austrian Academy of Sciences

# NGC 1817: the richest population of $\delta$ Scuti stars

M. F. Andersen[1], T. Arentoft[1], S. Frandsen[1],
L. Glowienka[1], H. R. Jensen[1], and F. Grundahl[1]

[1] Department of Physics and Astronomy, Aarhus University,
Ny Munkegade 120, Bldg. 1520, 8000 Århus C., Denmark

## Abstract

We combined two sets of data to get high-precision time-series for stars in the open cluster NGC 1817. We verified most of the variable stars detected earlier and found some new ones. To the 12 $\delta$ Scuti variables detected prior to this article 4 new $\delta$ Scuti stars and one candidate have been added. In addition 3 $\gamma$ Doradus candidates have been detected. The total number of variables and possible variables in NGC 1817, or in the field of view of the cluster is now up to 26. This high number of pulsating stars and especially $\delta$ Scuti variables makes this cluster an interesting asteroseismic target for future study. Using asteroseismology on $\delta$ Scuti stars will hopefully provide us with information that eventually will constrain the theoretical models of stars. This will in the end give us a much greater understanding of the evolution of stellar objects.

Accepted: 2009, July 8
Individual Objects: NGC 1817

## 1. Introduction

Asteroseismic study of pulsating stars is a very interesting field in astronomy nowadays. Results from asteroseismic investigation of oscillation modes of variable stars are helping us to understand fundamental parameters of stars. $\delta$ Scuti variable stars are stars in the range from A to F types and with masses ranging from 1.5 $M_\odot$ to 2.5 $M_\odot$. These stars are main sequence or post-main sequence stars and lie at the bottom of the classical Cephei instability strip. $\delta$ Scuti stars show multiperiodic signals with periods of the order of hours and magnitude variation at the mmag level. Through asteroseismology it is possible to constrain the stellar models, and to test the stellar physics to higher accuracy.

In the open cluster NGC 1817 different kinds of variable stars are present. A high number of $\delta$ Scuti stars and at least 2 eclipsing binaries make this cluster a prominent target for asteroseismic investigation. Observations of the area including NGC 1817 made detection of at least 16 $\delta$ Scuti variables possible. 12 of these were detected prior to this article (Frandsen & Arentoft 1998 and Arentoft et al. 2005) but the last 4 are new detections. Another 4 stars are possible candidates. Open clusters are interesting targets for CCD photometry, because of the semicrowded field, which make precise time-series photometry for the individual stars possible. For NGC 1817 the turnoff from the main sequence, due to the exhaustion of Hydrogen in the core of the stars, is located inside the instability strip. This makes NGC 1817 an obvious choice for CCD observations, because the stars in the instability strip, the pulsating candidates, are among the brightest stars in the cluster.

## 2.   Observations

Two datasets were used for the analysis of variable stars in the open cluster NGC 1817. The first set of data was taken with the 1.5 m Danish Telescope at La Silla, Chile. The observations were carried out by Lars Glowienka and Henrik Robenhagen Jensen in January 2005. The DFOSC instrument was used to obtain time-series CCD images. A 2048 x 2048 CCD camera was used for the I, B and V filter data. The main part of the total observations were carried out using the B filter and a small part with the V and I filter. In Fig. 1 an image from the La Silla data is shown with the detected variables indicated.

The second dataset was collected using the 2.56 m Nordic Optical Telescope (NOT) at La Palma, Canary Islands. These observations were performed in January 2007 by Lars Glowienka. A mosaic of four 1024 x 1024 CCD cameras were used, and to cover even more, three areas of the cluster were observed. Two images were taken in one area, then the telescope was pointed to a new area where two images were taken, and then pointed to the last area where two images were taken. Then it was moved back to the first area again and so on. One night (13-01-2007) the observations were done with the V filter. The last 3 nights (14,15,16-01-2007) were observed with the B filter.

### 2.1.   Data reduction

The two datasets were reduced separately using the Multi-Object Multi-Frame package (MOMF; Kjeldsen & Frandsen 1992). The package uses a point-spread function (PSF) and aperture photometry to determine the magnitude of every star in the observed field. In the La Silla data, time-series for 627 stars were obtained. The reduction of the data from NOT was a little more difficult. Because of drift in the position from night to night, every CCD area was reduced separately, which resulted in 12 different MOMF-reductions per

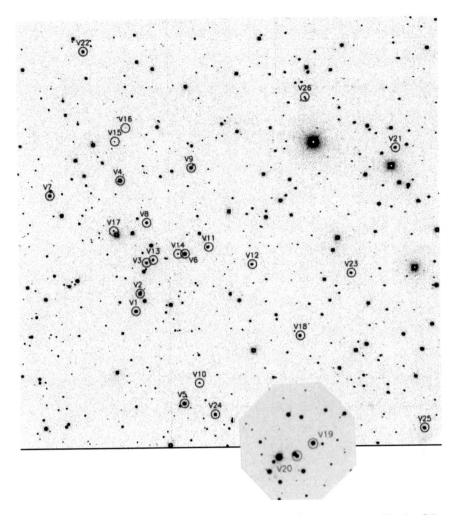

Figure 1: Finding chart for the variable stars (indicated) in NGC 1817. The La Silla field of view is ∼15'x15' and the inserted bottom image is part of one CCD image from NOT. It is rotated and scaled to match the La Silla image.

night. After the reduction, data for the detected variables were combined by normalizing the magnitude for each night. This affects the analysis of the possible γ Doradus stars, because of their long variation periods. In the NOT data around 60 stars were present in all of the 12 areas. Time-series of around 650 stars were obtained from the NOT data, because some of the areas were overlapping. Using MOMF, a precision of ∼2 mmag for the brightest stars was obtained for the La Silla data, and ∼1 mmag for the NOT data.

# 3.   Results

## 3.1.   Detecting the variables

The analysis of the time-series was done manually. Every star was checked visually for variability and the interesting stars were analyzed with the program Period04 (Lenz & Breger 2005). The investigation was done with the B filter data.

In Fig. 2, a part of the light curves for the new variable stars in NGC 1817 and some candidate variables in the field is shown. The frequency solution for the stars is superimposed with a solid line. These frequencies were determined with Period04, which applies a least square fit of sinusoidals to the data. A criteria of a signal-to-noise ratio (S/N) of 4 for the detected frequencies was used (Breger et al. 1993 & Kuschnig et al. 1997). Because of the window-function this criteria is not sufficient, but gives a rough estimate of the range and number of oscillations present in the stars. Long period variations were subtracted from the data when it was clear that the variations were not intrinsic. In Table 1 basic data for all the variable stars are shown, including candidates. The first 18 variables were detected by Frandsen & Arentoft (1998) and Arentoft et al. (2005).

The V and B-V magnitudes for the variable V20 in Table 1 were determined by comparison of 2 stars in the overlapping region for the La Silla data and the NOT data. Then the transformation was applied to V20, which was done to get an idea about the position of the star in the color-magnitude diagram (CMD). The transformation will be explained in more detail below.

## 3.2.   Frequency analysis

In Table 2 the frequency solution for the new variables is listed. Only the stars for which a significant solution could be determined are shown. Oscillations with a S/N higher than 4 were used. Because of the low number of detected frequencies, the S/N of every solution is listed as well. The noise level was determined after removing the signals from the table.

In Fig. 2 the light curves for the new variables are shown. The solid line is the least square fit solution, with the frequencies from Table 2. Because of the reduction technique no frequency analysis was possible for the two $\gamma$ Doradus variables V23 and V24. The vertical dashed lines in these two plots are to indicate that it is not possible to combine the light curve from the different nights. Therefore no frequency analysis of these stars is possible. V21 is clearly a multiperiodic $\delta$ Scuti star (cf. Fig. 2), but the data was not sufficient for a frequency solution.

Table 1: Basic data for the variable stars in NGC 1817

| ID | WIYN-ID | R.A. (2000.0) | Decl. (2000.0) | B | B-V | Class |
|----|---------|---------------|----------------|------|------|-------|
| V1 | 549 | 5 12 42.8 | 16 41 43 | 13.50 | 0.36 | $\delta$-Scuti |
| V2 | 641 | 5 12 40.8 | 16 42 00 | 12.85 | 0.41 | $\delta$-Scuti |
| V3 | 788 | 5 12 37.4 | 16 42 31 | 14.35 | 0.45 | $\delta$-Scuti |
| V4 | 985 | 5 12 32.2 | 16 44 52 | 12.58 | 0.41 | $\delta$-Scuti / Binary |
| V5 | 386 | 5 12 46.8 | 16 38 40 | 12.86 | 0.42 | $\delta$-Scuti |
| V6 | 963 | 5 12 33.0 | 16 41 50 | 12.90 | 0.44 | $\delta$-Scuti |
| V7 | 650 | 5 12 40.1 | 16 46 07 | 13.71 | 0.39 | $\delta$-Scuti |
| V8 | 939 | 5 12 33.7 | 16 43 22 | 14.34 | 0.47 | $\delta$-Scuti |
| V9 | 1331 | 5 12 24.6 | 16 43 32 | 13.18 | 0.49 | $\delta$-Scuti |
| V10 | 534 | 5 12 43.6 | 16 38 46 | 16.28 | 0.46 | $\delta$-Scuti |
| V11 | 1090 | 5 12 30.3 | 16 41 28 | 14.28 | 0.44 | $\delta$-Scuti |
| V12 | 1203 | 5 12 27.8 | 16 40 07 | 14.64 | 0.50 | $\delta$-Scuti |
| V13 | 823 | 5 12 36.5 | 16 42 25 | 14.68 | 0.49 | ... |
| V14 | 943 | 5 12 33.7 | 16 42 00 | 16.40 | 0.87 | ... |
| V15 | 1116 | 5 12 29.1 | 16 45 49 | 18.23 | 1.00 | ... |
| V16 | - | 5 12 26.9 | 16 45 52 | 18.91 | 1.13 | Binary |
| V17 | 773 | 5 12 37.4 | 16 43 56 | 16.63 | 0.72 | $\delta$-Scuti |
| V18 | 1128 | 5 12 29.9 | 16 37 30 | 14.17 | 0.47 | Binary |
| V19 | - | 5 12 38.4 | 16 34 54 | 13.28 | 0.33 | $\delta$-Scuti |
| V20 | - | 5 12 41.2 | 16 34 58 | $\sim$ 14.24 | $\sim$ 0.57 | $\delta$-Scuti |
| V21 | - | 5 12 04.0 | 16 39 21 | 14.37 | 0.42 | $\delta$-Scuti |
| V22 | - | 5 12 23.8 | 16 48 27 | 13.97 | 0.47 | $\delta$-Scuti |
| V23 | - | 5 12 19.5 | 16 37 41 | 14.64 | 0.46 | $\gamma$-Dor |
| V24 | 476 | 5 12 45.0 | 16 37 44 | 13.41 | 0.80 | $\gamma$-Dor |
| V25 | - | 5 12 26.9 | 16 32 42 | 13.28 | 0.67 | $\gamma$-Dor |
| V26 | - | 5 12 07.6 | 16 42 29 | 16.81 | 0.69 | $\delta$-Scuti |

The variable ID for the first 18 stars is that given by Arentoft et al. (2005) and the WIYN-ID is the star number from the WIYN catalog. Right ascension is given in hours, minutes and seconds and declination is given in degrees, arcminutes and arcseconds.

Table 2: Pulsational data for the new variables in NGC 1817

| ID | $f_1$ | $(S/N)_1$ | $f_2$ | $(S/N)_2$ | $f_3$ | $(S/N)_3$ | Class |
|----|-------|-----------|-------|-----------|-------|-----------|-------|
| V19 | 10.89 | 25.7 | 11.73 | 21.7 | 14.83 | 4.57 | $\delta$-Scuti |
| V20 | 24.05 | 5.21 | ... | ... | ... | ... | $\delta$-Scuti |
| V22 | 14.15 | 9.14 | ... | ... | ... | ... | $\delta$-Scuti |
| V25 | 0.3507 | 130 | ... | ... | ... | ... | $\gamma$-Dor |
| V26 | 4.653 | 5.80 | 6.112 | 4.90 | ... | ... | $\delta$-Scuti |

The frequencies are given in cycles per day and the S/N ratio is shown for every frequency.

## 3.3. The color-magnitude diagram

A transformation from the observed V and B data, from La Silla, was done using the WIYN Open Cluster Study (Wisconsin, Indiana, Yale and NOAO) catalog of standardized CCD photometry. 96 stars, which were identified in both WIYN and the La Silla data, were used to obtain the offset transformation. The

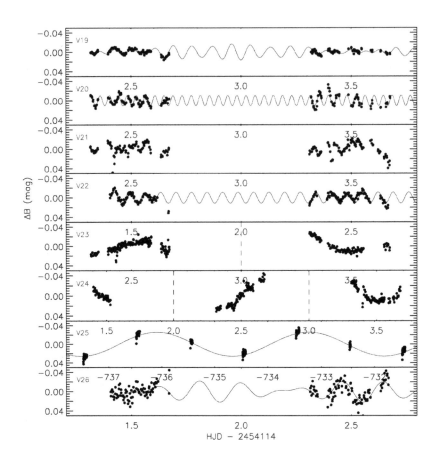

Figure 2: Light curves for the 8 new variables

position of the stars was used to match the different stars in the two data sets.

In Fig. 3 the CMD for the stars in the observed field from La Silla is shown. The variable stars are indicated; diamonds indicate δ Scuti variables, squares indicate binary systems, triangles indicate γ Doradus variables and circles indicate candidate variables.

The instability strip (Breger 2000) for NGC 1817 is superimposed with the dashed lines. It is evident that most of the variables are located inside the instability strip. We also see that the stars inside the strip are positioned close to the cluster sequence, which means that most of the stars in the instability strip are likely cluster members. Errors in the CMD are dominated by transformation errors, but the figure is in good agreement with the CMD determined

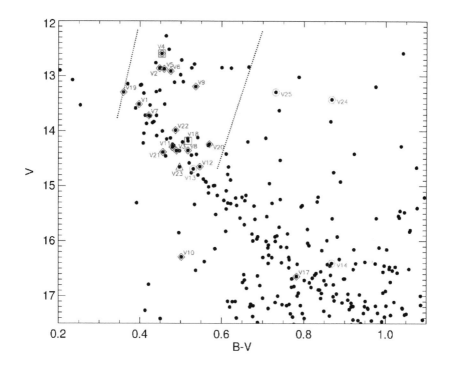

Figure 3: Color-Magnitude diagram for stars in i NGC 1817

by Arentoft et al. (2005).

## 4. Conclusion

The detection of 17 $\delta$ Scuti variables, 2 eclipsing binaries (one with a component that is also a $\delta$ Scuti) and 8 other possible variables, 4 which could be $\gamma$ Doradus variables, one contact binary, one $\delta$ Scuti and 2 other candidates, is at the moment the highest concentration of $\delta$ Scuti variables in any open cluster. This makes NGC 1817 a very interesting asteroseismic target. The combination of $\delta$ Scuti stars and eclipsing binary systems in an open cluster makes this particular target of greatest interest.

Membership determination of the variables by considering the proper motion, and spectroscopic study of the eclipsing binaries will be the next step. J. Molenda-Żakowitz et al. (2009) used spectroscopic observations of 11 variable stars in NGC 1817 to analyze these stars further. In the future a multisite campaign will be needed to get sufficient time-series data to make asteroseismic study of the individual stars possible.

## References

Arentoft, T., Bouzid, M. Y., Sterken, C. et al. 2005, PASP, 117, 601

Balaguer-Núñez, L., Jordi, C., Galadí-Enríquez, D., et al. 2008, A&A, 426, 819

Breger, M., Stich, J., Garrido, R. et al. 1993, A&A, 271, 482

Breger, M. 2000, ASPC, 210, Proceedings of Delta Scuti and Related Stars, ed. M. Breger & M. H. Montgomery (San Francisco: ASP), 3

Frandsen, S., & Arentoft, T. 1998, A&A, 333, 524

Kjeldsen, H., & Frandsen, S. 1992, PASP, 104, 413

Kuschnig, R., Weiss, W. W., Gruber, R., et al. 1997, A&A, 328, 544

Lenz, P., & Breger, M. 2005, CoAst, 146, 53

Machado, L. Fox., Schuster, W. J., Zurita, C., et al. 2008, CoAst, 156, 27

Molenda-Żakowitz, J., Arentoft, T., Frandsen, S., et al. 2009, AcA, 59, 69

Comm. in Asteroseismology
Volume 160, August 2009
© Austrian Academy of Sciences

# The frequency spectrum of periodically modulated sinusoidal oscillation

B. Szeidl, and J. Jurcsik

Konkoly Observatory of the Hungarian Academy of Sciences.
P.O. Box 67, H-1525 Budapest, Hungary

## Abstract

The mathematical model of periodically amplitude and phase modulated sinusoidal oscillation is studied, and its Fourier spectrum is given analytically. The Fourier spectrum of the model explains the main features of the frequency spectrum of RR Lyrae stars showing light curve modulation called the Blazhko effect: among others the appearance of multiplets, the rapid decrease of their amplitudes in increasing orders, the asymmetry of the amplitudes of the side frequency pairs, and the possibility of the occurrence of frequency doublets instead of triplets in the spectrum. The good agreement of the results of this mathematical model with observational facts favours those physical models of the Blazhko effect which explain the phenomenon as a modulation of the oscillation with the modulation frequency, $f_m$.

Accepted:     2009, August 19

## 1.   Introduction

In the past years contradictory descriptions have been published about the frequency spectrum of the RR Lyrae stars with modulated light curves. Analyzing the light curves of RRab stars in the large data bases of MACHO and OGLE (Alcock et al. 2003; Moskalik & Poretti 2003; Collinge et al. 2006) different classifications of the modulation were introduced depending on the number and the separation of the modulation side-frequencies detected in the spectrum. In variables with an equidistant triplet structure around the fundamental mode frequency, no further multiplet structure could be identified in these studies. The observations focused on individual Blazhko stars, however, inevitably proved the presence of higher multiplets (quintuplet, septuplet structures) in the spectrum (Hurta et al. 2008; Jurcsik et al. 2008; Kolenberg et al. 2009, Jurcsik et al. 2009a). Up to now the only mathematical model that aimed to describe the full spectrum with multiplets of Blazhko variables was published by Breger & Kolenberg (2006). However, no detailed confrontation with the observations of the possible predictions of this model e.g., on the amplitudes of the components of the multiplets has been performed.

As the frequency spectrum implies the basic information (amplitudes and phase angles) on the light curve modulation, its correct knowledge and interpretation is very important for proper understanding of the Blazhko effect. If we find the adequate mathematical model that describes all the features of the frequency spectrum of the well-observed Blazhko stars (frequencies, amplitudes and phase relations), it may provide a starting-point and steady base for theoretical investigations.

## 2.  The frequency spectrum of periodically modulated sinusoidal oscillation

The amplitude and phase modulated sinusoidal oscillation is given by the formula

$$m(t) = a\left[1 + b\,\sin\left(\Omega t + \varphi_1\right)\right]\sin\left[\omega t + \varphi_0 + c\,\sin\left(\Omega t + \varphi_2\right)\right], \qquad (1)$$

where $\omega = 2\pi f_0$, $f_0$ is the fundamental frequency, $\Omega = 2\pi f_m$, $f_m$ is the frequency of the modulation, $a$, $b$ and $c = 2\pi f_0 q$ are the amplitudes of the oscillation, the amplitude and the phase modulations, respectively. $q$ expresses the amplitude of the phase modulation relative to the fundamental period, $\varphi_0$, $\varphi_1$ and $\varphi_2$ denote the phases of the oscillation, amplitude and phase modulations, respectively. Analytically, there is no limit for the parameters $a$, $b$ and $c(q)$, but because of observational constraints we confine our discussion to the parameter values $0 \le b \le 1$ and $|c| \le \pi/2$. Solutions for parameters out of these ranges are irrelevant in asteroseismology.

We do not consider modulation of the oscillation frequency here as the modulated oscillation of the form of

$$m(t) = a\,\sin[\omega(1 + d\sin(\Omega + \varphi_3))t + \varphi_0] =$$
$$a\,\sin[\omega t + \varphi_0 + d\omega t\sin(\Omega t + \varphi_3)]$$

corresponds to a phase modulation with variable modulation amplitude ($c = d\omega t$) and time dependent, unstable frequency spectrum.

By suitable choice of the starting epoch, without any restriction on the general validity, $\varphi_0 = 0$ and $\varphi_2 = 0$ can be attained. If the initial epoch corresponds to the timing of both the mid rising branch of the phase modulation and the mid descending branch of the oscillation (taking into account the reverse direction of the magnitude scale) both $\varphi_0 = 0$ and $\varphi_2 = 0$ fulfill. Since $f_0$ and $f_m$ can be regarded as rational numbers as the accuracy of their numerical value is limited by the observations, such an epoch should exist. We note here that with cosine representation the choice of the appropriate initial epoch would be more natural, it would correspond to the maxima of the oscillation and the phase modulation.

Denote now $\Phi = \varphi_1 - \varphi_2$ the epoch independent phase difference between the amplitude and phase modulations. Then the time history of the modulated oscillation is described as follows

$$m(t) = a\left[1 + b\sin\left(\Omega t + \Phi\right)\right]\sin\left(\omega t + c\sin\Omega t\right). \tag{2}$$

Taking the simple trigonometric addition formula of $\sin(\alpha + \beta)$ and the Taylor-series of $\sin x$ and $\cos x$ into account, Eq. 2 has the form

$$m(t) = a\left(1 + b\sin\Phi\cos\Omega t + b\cos\Phi\sin\Omega t\right) \cdot$$

$$\left[\sin\omega t\sum_{n=0}^{\infty}(-1)^n\frac{c^{2n}}{(2n)!}\sin^{2n}\Omega t + \cos\omega t\sum_{n=0}^{\infty}(-1)^n\frac{c^{2n+1}}{(2n+1)!}\sin^{2n+1}\Omega t\right] \tag{3}$$

Substituting the well-known power-reduction formulae Eqs. 4 and 5 into Eq. 3 with $\alpha = \Omega t$

$$\sin^{2n}\alpha = \frac{1}{2^{2n-1}}\left[-\frac{1}{2}\binom{2n}{n} + \sum_{k=0}^{n}(-1)^k\binom{2n}{n-k}\cos 2k\alpha\right] \tag{4}$$

and

$$\sin^{2n+1}\alpha = \frac{1}{2^{2n}}\sum_{k=0}^{n}(-1)^k\binom{2n+1}{n-k}\sin(2k+1)\alpha \tag{5}$$

and then applying the simple trigonometric formulae

$$\sin\alpha\sin\beta = \frac{1}{2}\left[\cos\left(\alpha - \beta\right) - \cos\left(\alpha + \beta\right)\right]$$

$$\sin\alpha\cos\beta = \frac{1}{2}\left[\sin\left(\alpha - \beta\right) + \sin\left(\alpha + \beta\right)\right] \tag{6}$$

$$\cos\alpha\cos\beta = \frac{1}{2}\left[\cos\left(\alpha - \beta\right) + \cos\left(\alpha + \beta\right)\right]$$

where $\alpha = \omega t$, $\beta = i\Omega t$, $i = 1, 2, 3, \ldots$, Eq. 3 will have the form

$$
\begin{aligned}
m(t) = x_0 \sin \omega t + z_0 \cos \omega t &+ \sum_{i=1}^{\infty} x_i \sin(\omega + i\Omega)t + \sum_{i=1}^{\infty} y_i \sin(\omega - i\Omega)t \\
&+ \sum_{i=1}^{\infty} z_i \cos(\omega + i\Omega)t + \sum_{i=1}^{\infty} w_i \cos(\omega - i\Omega)t
\end{aligned}
\tag{7}
$$

where $x_i$, $y_i$, $z_i$ and $w_i$ ($i = 1, 2, 3, \ldots$) coefficients depend only on $a$, $b$, $c$ and $\Phi$. Applying the known relations Eqs. 8 and 9 to Eq. 7:

$$
\cos \alpha = \sin \left( \alpha + \frac{\pi}{2} \right), \qquad \alpha = (\omega \pm i\Omega)\, t \quad i = 0, 1, 2, \ldots
\tag{8}
$$

and

$$
X_1 \sin \nu t + X_2 \sin \left( \nu t + \frac{\pi}{4} \right) = X \sin (\nu t + \varphi), \qquad \nu = \omega \pm i\Omega \quad (i = 0, 1, 2, \ldots)
\tag{9}
$$

where $X^2 = X_1^2 + X_2^2$ and $\tan \varphi = X_1/X_2$ we arrive at the Fourier spectrum of the modulated sinusoidal oscillation:

$$
\begin{aligned}
m(t) = A_0 \sin(\omega t + \chi_0) &+ \sum_{i=1}^{\infty} A_i^+ \sin \left[ (\omega + i\Omega)\, t + \chi_i^+ \right] \\
&+ \sum_{i=1}^{\infty} A_i^- \sin \left[ (\omega - i\Omega)\, t + \chi_i^- \right]
\end{aligned}
\tag{10}
$$

By using the outlined procedure we can derive the $A_0$, $A_i^+$ and $A_i^-$ coefficients (amplitudes) as well as their phase angles, $\chi_0$, $\chi_i^+$ and $\chi_i^-$.

As an example we derive the amplitudes and phases of the triplet. In this case only the $k = 0, 1$ terms in Eq. 4 and the $k = 0$ term in Eq. 5 should be considered and substituted into Eq. 3 for each $n$ ($n = 0, 1, 2, \ldots$). All the other terms contribute to the higher members of the multiplets. In this case taking Eqs. 6 into account, Eq. 3 takes the form:

$$
\begin{aligned}
m(t) = a\,(1 &+ b \sin \Phi \cos \Omega t + b \cos \Phi \sin \Omega t) \cdot \\
&\cdot \left\{ \sin \omega t \sum_{n=0}^{\infty} (-1)^n \frac{c^{2n}}{2^{2n-1}(2n)!} \left[ \binom{2n}{n} - \binom{2n}{n-1} \cos 2\Omega t - \frac{1}{2} \binom{2n}{n} \right] \right. \\
&\left. + \cos \omega t \sum_{n=0}^{\infty} (-1)^n \frac{c^{2n+1}}{2^{2n}(2n+1)!} \binom{2n+1}{n} \sin \Omega t \right\} + g
\end{aligned}
\tag{11}
$$

where $g$ is a function of $\omega \pm k\Omega$, $k \geq 2$.

Let the convergent series

$$S = \sum_{n=0}^{\infty} (-1)^n \left( \frac{c^n}{2^n n!} \right)^2 = 1 - \frac{c^2}{4} + \frac{c^4}{64} - \frac{c^6}{2304} + \frac{c^8}{147456} - \frac{c^{10}}{14745600} + \dots \quad (12)$$

and

$$S_1 = \sum_{n=0}^{\infty} (-1)^n \left( \frac{c^n}{2^n n!} \right)^2 \frac{n}{n+1} = -\frac{c^2}{8} + \frac{c^4}{96} - \frac{c^6}{3072} + \frac{c^8}{184320} - \frac{c^{10}}{17694720} + \dots$$

$$(13)$$

Since

$$\binom{2n}{n} = \frac{(2n)!}{(n!)^2}; \quad \binom{2n}{n-1} = \frac{(2n)!}{(n!)^2} \frac{n}{n+1}; \quad \binom{2n+1}{n} = \frac{(2n+1)!}{(n!)^2} \frac{1}{n+1}$$

Eq. 11 takes the form:

$$m(t) = a \left( 1 + b \sin \Phi \cos \Omega t + b \cos \Phi \sin \Omega t \right) \cdot$$

$$\cdot \left[ S \sin \omega t - 2 S_1 \sin \omega t \cos 2\Omega t + (S - S_1) c \cos \omega t \sin \Omega t \right] + g \quad (14)$$

If we execute the multiplications in Eq. 14 taking into account Eqs. 6, and disregard all the terms with $\omega \pm k\Omega$, $k \geq 2$, we obtain the following equation that describes the triplet structure:

$$m(t)_{\text{triplet}} = a \left\{ \left[ S \sin \omega t + \frac{1}{2}(S - S_1) bc \cos \Phi \cos \omega t \right] \right.$$

$$+ \left[ \frac{1}{2} Sb \sin \Phi - \frac{1}{2} S_1 b \sin \Phi - \frac{1}{2}(S - S_1)c \right] \sin(\omega - \Omega)t$$

$$+ \left[ \frac{1}{2} Sb \cos \Phi + \frac{1}{2} S_1 b \cos \Phi \right] \cos(\omega - \Omega)t \quad (15)$$

$$+ \left[ \frac{1}{2} Sb \sin \Phi - \frac{1}{2} S_1 b \sin \Phi + \frac{1}{2}(S - S_1)c \right] \sin(\omega + \Omega)t$$

$$+ \left. \left[ -\frac{1}{2} Sb \cos \Phi + \frac{1}{2} S_1 b \cos \Phi \right] \cos(\omega + \Omega)t \right\}$$

Application of Eqs. 8 and 9 to Eq. 15 yields the squares of the amplitudes as well as the phases of the triplet:

$$A_0^2 = a^2 [S^2 + \frac{1}{4}(S - S_1)^2 b^2 c^2 \cos^2 \Phi] \quad (16)$$

$$(A_1^+)^2 = a^2 \left[ \frac{1}{4}(S - S_1)^2(b^2 + c^2 + 2bc \sin \Phi) + SS_1 b^2 \cos^2 \Phi \right]$$

$$(A_1^-)^2 = a^2 \left[ \frac{1}{4}(S - S_1)^2(b^2 + c^2 - 2bc \sin \Phi) + SS_1 b^2 \cos^2 \Phi \right]$$

(17)

$$\tan \chi_0 = \frac{1}{2} \frac{S - S_1}{S} bc \cos \Phi \tag{18}$$

$$\tan \chi_1^+ = -\frac{(S + S_1)b \cos \Phi}{(S - S_1)(b \sin \Phi + c)}$$

$$\tan \chi_1^- = -\frac{(S + S_1)b \cos \Phi}{(S - S_1)(b \sin \Phi - c)}$$

(19)

Note that an interesting relation exists for the power difference of the side frequencies:

$$(A_1^+)^2 - (A_1^-)^2 = (S - S_1)^2 a^2 bc \sin \Phi \tag{20}$$

Eq. 20 shows that the asymmetry of the side frequency amplitudes depends basically on the phase difference between the amplitude and phase modulation components.

The amplitudes and phases of the subsequent frequencies of the multiplets can be derived in the same way. E.g., the final result for the side frequencies of the quintuplet ($k = 2$) is as follows. Let the convergent series

$$S_2 = \sum_{n=0}^{\infty} (-1)^n \left( \frac{c^n}{2^n n!} \right)^2 \frac{n}{(n+1)(n+2)}$$

$$= -\frac{c^2}{24} + \frac{c^4}{384} - \frac{c^6}{15360} + \frac{c^8}{1105920} - \frac{c^{10}}{123863040} + \dots$$

then

$$(A_2^+)^2 = a^2 \left[ S_1^2 - \frac{1}{2}(S - S_1 - S_2)bc \sin \Phi \right.$$

$$\left. + \frac{1}{16}(S - S_1 - S_2)^2 b^2 c^2 + \frac{1}{4}S_2(S - S_1)b^2 c^2 \cos^2 \Phi \right]$$

and

$$(A_2^-)^2 = a^2 \left[ S_1^2 + \frac{1}{2}(S - S_1 - S_2)bc \sin \Phi \right.$$

$$\left. + \frac{1}{16}(S - S_1 - S_2)^2 b^2 c^2 + \frac{1}{4}S_2(S - S_1)b^2 c^2 \cos^2 \Phi \right]$$

while

$$\tan \chi_2^+ = \frac{(S - S_1 + S_2)bc \cos \Phi}{4S_1 - (S - S_1 - S_2)bc \sin \Phi}$$

$$\tan \chi_2^- = \frac{(S - S_1 + S_2)bc\cos\Phi}{4S_1 + (S - S_1 - S_2)bc\sin\Phi}.$$

Note again that the power difference of the frequency pair simply depends on $\sin\Phi$:

$$(A_2^+)^2 - (A_2^-)^2 = -(S - S_1 - S_2)a^2 bc\sin\Phi.$$

## 3. Discussion

In realistic cases, in Blazhko stars the harmonics of the fundamental frequency are also present in the oscillation and appear in the frequency spectrum. So the time history of the periodically modulated oscillation is described by the equation

$$m(t) = \sum_{i=1}^{\infty} a_i \left[1 + b_i \sin\left(\Omega t + \Phi_i\right)\right] \sin\left[i\omega t + c_i \sin\Omega t\right]. \tag{21}$$

In the previous section we confined ourselves to the fundamental frequency. The same deduction is, however, valid and true for the harmonics, simply $\omega$ should be replaced by $i\omega$. Therefore, we conclude that the frequency spectrum of any periodically modulated periodic oscillation i.e. the light curve of Blazhko stars is the infinite series of the multiplets of the fundamental frequency and its harmonics:

$$m(t) = \sum_{i=1}^{\infty} A_{i0}\sin(i\omega t + \chi_{i0}) + \sum_{i=1}^{\infty}\sum_{j=1}^{\infty} A_{ij}^+ \sin\left[(i\omega + j\Omega)t + \chi_{ij}^+\right]$$
$$+ \sum_{i=1}^{\infty}\sum_{j=1}^{\infty} A_{ij}^- \sin\left[(i\omega - j\Omega)t + \chi_{ij}^-\right] \tag{22}$$

We also note here that, if the sums in Eq. 22 start from $i = 0$, then this description gives a natural ground to the appearance of $A_{00}$, and frequencies at $jf_m$, which correspond to an arbitrary zero point of the scale, and the modulation frequency and its harmonics, respectively. Although the appearance of $f_m$ (and perhaps $2f_m$) is hardly (if at all) discernible in most of the observed frequency spectra of Blazhko stars, in accurate, extended data the modulation frequency can be always detected with very small amplitude. This is most probably due to the fact that the star's physical parameters (e.g. mean luminosity, temperature) change only weakly during the Blazhko cycle (Jurcsik et al. 2009a, 2009b).

A priori there is no definite connection between the parameters of the modulation. The rough constancy of the phase differences and the systematic behaviour of the amplitudes of the triplet components with increasing orders $i$

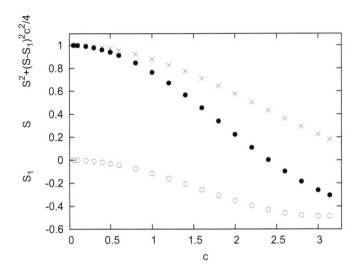

Figure 1: Variations of the $S$ (dots), $S_1$ (circles) sums and their combination term (crosses) which define the amplitude decrease of the $f_0$ frequency component with increasing phase modulation amplitude, $c$.

(see Fig. 10 in Jurcsik et al. 2009a) suggests, however, the existence of some connection between them. If we suppose that $b$, $c$ and $\Phi$ are unique parameters of the modulation in the different harmonic orders then $A_{ij}^{+/-}$ and $\chi_{ij}^{+/-}$ would depend only on $a_i$. In this case the side frequencies of the different order pulsation components would naturally have some common properties.

In all probability the amplitudes of the $if_0$ frequencies in the spectrum of modulated light curves run more or less similarly to that of the single periodic RR Lyraes. It should, however, be noted that according to Eq. 16, $a^2[S^2 + (S - S_1)^2 c^2/4]$ is an upper limit ($b = 1, \Phi = 0$) for the amplitude of $f_0$, i.e., $A_0/a = 1$ only if the modulation is pure amplitude modulation ($c = 0$). Any phase modulation component lowers the amplitude of the $f_0$ frequency. Fig. 1 shows the dependence of $S, S_1$ and the upper limit of the amplitude reduction factor ($S^2 + (S - S_1)^2 c^2/4$) on $c$, using $f_0 = 2.1925$ c/d oscillation frequency, corresponding to the pulsation frequency of DM Cyg. The phase modulation lowers the amplitudes of the harmonic components, too, ($A_{i0}/a_i < 1$ if $c_i > 0$) as E.q. 16 also holds for the frequencies $\omega = i\omega$.

As mentioned above, the accurate and well-distributed (over both frequencies $f_0$ and $f_m$) observations of Blazhko stars clearly show the multiplet structure around the fundamental mode frequency and its harmonics in their Fourier spectrum (Jurcsik et al. 2008, Kolenberg et al. 2009). The question arises why the triplet is a striking feature while the higher multiplets are hardly perceptible in the frequency spectrum of a Blazhko star.

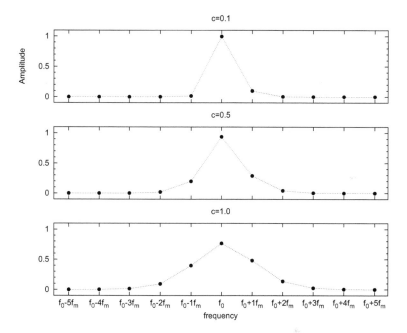

Figure 2: Amplitude decrease of the multiplet components in the different orders of the modulation. Artificial datasets were generated for different values of $c$ (0.1, 0.5, 1.0) according to Eq. 3 using $a = 1$, $b = 0.1$ and $\Phi = 72°$ values and $f_0, f_m$ frequencies of DM Cyg. The phase modulation with $c = 1$ amplitude corresponds to full amplitude of the phase modulation slightly less than $1/3$ of the main period which is substantially larger than the amplitude of the phase modulation observed in Blazhko stars, which is typically about $0.1 - 0.2$ pulsation phase. Even in the $c = 1$ simulations the amplitudes of the higher order modulation components ($f_0 \pm kf_m$, $k \geq 3$) are neglectably small.

Although the formulae of the amplitudes of the multiplets can be exactly derived as it was shown in the previous section, they are fairly complicated expressions (besides the outlined procedure is too laborious). Their behaviour can be, however, easily studied through the Fourier analysis of synthetic light curves. As examples, Figs. 2, 3 and Table 1 show the amplitudes of subsequent multiplets of the frequency spectra of Eq. 2 at different choice of the modulation

Table 1: Amplitudes of the frequency multiplets appearing in the Fourier spectrum of amplitude and phase modulated sinusoidal signal according to Eq. 2 using $a = 1, b = 0.1, c = 0.5$ parameters for different $\Phi$ values.

| $\Phi$ | $f_0 - 5f_m$ | $f_0 - 4f_m$ | $f_0 - 3f_m$ | $f_0 - 2f_m$ | $f_0 - f_m$ | $f_0$ |
|---|---|---|---|---|---|---|
| 36° | 0.000007 | 0.000134 | 0.002067 | 0.025337 | 0.216919 | 0.938674 |
| 72° | 0.000003 | 0.000055 | 0.001197 | 0.019320 | 0.196687 | 0.938500 |
| 90° | 0.000000 | 0.000032 | 0.001025 | 0.018363 | 0.193815 | 0.938470 |

| $\Phi$ | $f_0 + 5f_m$ | $f_0 + 4f_m$ | $f_0 + 3f_m$ | $f_0 + 2f_m$ | $f_0 + f_m$ |
|---|---|---|---|---|---|
| 36° | 0.000014 | 0.000258 | 0.003680 | 0.039023 | 0.273228 |
| 72° | 0.000016 | 0.000286 | 0.004054 | 0.042408 | 0.288692 |
| 90° | 0.000016 | 0.000289 | 0.004102 | 0.042845 | 0.290722 |

parameters. The striking feature is that the amplitudes of the subsequent frequencies on both sides of the fundamental frequency approach zero rapidly. The degree of the decrease depends on the parameter of the phase modulation $c$, the weaker the phase modulation, the faster the amplitude decrease is. Even if the phase modulation has a high value, $c = 1$, the amplitude of the $f_0 \pm 5f_m$ frequencies are more than three orders of magnitude less than the amplitude of $f_0$ (see Fig 2). For a more realistic case e.g., with $c = 0.5$ the amplitude difference between the first and fifth order modulation components is larger than four orders of magnitude as shown in Table 1. It is now clear that only the observational accuracy sets limit to the perception of higher multiplets in the frequency spectrum.

Another important implication of our results is that Eq. 20 proves that the power difference of the side frequencies in the triplet is the physically meaningful quantity to measure the asymmetry of the triplet instead of their amplitude ratios.

Our calculations have further serious and amazing consequences on the symmetry and asymmetry of the triplet components. It results also from Eq. 20 that the triplet will only be symmetrical if any of the quantities $b$, $c$ or $\Phi$ equals to zero, or $\Phi = \pi$. If the quantities $b$, $c$ and $\Phi$ have appropriate numerical values, one of the amplitudes of the triplet side frequencies $A_1^+$ or $A_1^-$ may become zero. The only solution of the $(A_1^+)^2 = 0$ or $(A_1^-)^2 = 0$ second order equations (Eqs. 17) is $b = c$ and $\Phi = 3\pi/2$ or $\Phi = \pi/2$. In this case the spectrum of the modulated oscillation is doublet instead of triplet. As an instructive example Fig. 3 shows the amplitude changes of the frequencies as a function of $\Phi$ for $a = 1, b = 0.1$ and $c = 0.1, 0.5, 1.0$. In Blazhko stars, usually there is a phase difference between the amplitude and phase modulation

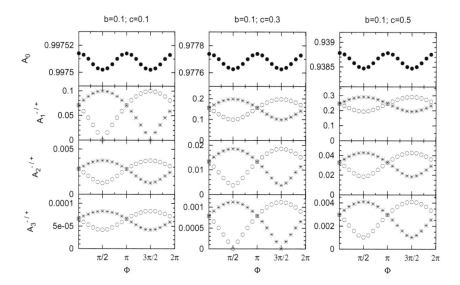

Figure 3: Amplitude variations of the multiplet frequency components of periodically amplitude and phase modulated sinusoidal signal as a function of $\Phi$, the phase difference between the amplitude and phase modulation components. $A_0, A_1^{+/-}, A_2^{+/-}, A_3^{+/-}$ denote the amplitudes of the $f_0$, $f_0 \pm f_m$, $f_0 \pm 2f_m$, and $f_0 \pm 3f_m$ frequency components. The three panels show the results for different combinations of $b$ and $c$, the amplitudes of the amplitude and phase modulations, respectively. Artificial datasets were generated using Eq. 2 with $a = 1$ and $f_0 = 2.3817$, $f_m = 0.06043$ c/d values corresponding to the pulsation and modulation frequencies of DM Cyg, and Fourier analysed in order to determine the amplitude values.

components (see e.g., Fig 8 in Jurcsik et al 2008, and Fig 3 in Jurcsik et al 2009a) and this explains why asymmetric triplets are detected in most of the cases.

Concerning the phases, from Eq. 18 it follows that if either $b = 0$ or $c = 0$ the phase of the oscillation is 0 or $\pi$ ($\tan \chi_0 = 0$) if the initial epoch is chosen as described in the beginning of Sect. 1. When the modulation is pure phase modulation, then $\tan \chi_1^- = \tan \chi_1^+ = 0$, i.e., $\chi_1^-$, and $\chi_1^+$ are 0 or $\pi$. When the modulation is pure amplitude modulation, then Eq. 1 can be easily solved with appropriate choice of the initial epoch. If the initial epoch is chosen to fulfill $\varphi_0 = 0$ and $\varphi_1 = 0$, i.e., it corresponds to the timing of both the mid rising branch of the amplitude modulation and the mid descending branch of the oscillation, the solution of Eq. 1 shows that $\chi_1^- = \pi/2$ and $\chi_1^+ = -\pi/2$.

Alternatively, if the initial epoch corresponds to the maximum phase of the amplitude modulation i.e., $\varphi_1 = \pi/2$, then $\chi_1^- = \chi_1^+ = \chi_0$ fulfill.

Eq. 20 provides another interesting result. If $0 < \Phi < \pi$, then $A_1^+ > A_1^-$ and the plot that shows the amplitude of the light variation vs. phase shift of maximum light during the Blazhko cycle has anticlockwise progression, while in the case of $-\pi < \Phi < 0$, $A_1^+ < A_1^-$, the progression is the opposite. In reality, of course, the situation is more complex because the modulations of the harmonic components may modify the picture.

## 4.  Conclusions

The model of periodically modulated harmonic oscillation properly explains the main features of the frequency spectrum of Blazhko RR Lyrae stars. It predicts the infinite multiplet systems around the fundamental frequency and its harmonics, accounts for the often detected asymmetry of the amplitudes of the side frequency pairs in the triplets (quintuplets) and for the rapid decrease of the amplitudes of the multiplets with increasing orders.

Our results also indicate that the suggested classification schemes of Blazhko stars based on their frequency spectrum (Alcock et al. 2003, Moskalik and Poretti 2003) are dubious. For example, the frequency doublets, which are generally interpreted with a nonradial frequency component close to the radial frequency, can naturally take its origin from amplitude and phase modulations of a single oscillation. Our analysis also shows that the asymmetry of the triplets are a natural consequence of the mixture of amplitude and phase modulations. We thus conclude that the occurrence of multiplets does not necessarily imply the presence of more than a single oscillation, and prefers those physical models which connect the phenomenon to one modulation frequency, $f_m$. Recently, Stothers (2006) proposed such a model for Blazhko RR Lyrae stars. In his model the cyclic changes in the strength of the envelope convection gives rise to the modulation of the periodic oscillation of these stars.

In the near future we plan further investigations to exploit the potential of this model and to find connections between the amplitudes and phase angles in the frequency spectrum. We also plan to carry out a more profound comparison with observational facts.

**Acknowledgments.** We thank the referee, Katrien Kolenberg for her helpful comments. We are grateful to Ádám Sódor for his help in preparing the manuscript. The financial support of OTKA grant T-068626 is acknowledged.

# References

Alcock, C., Alves, D. R., Becker, A., et al. 2003, ApJ, 598, 597

Breger, M., & Kolenberg, K. 2006, A&A 460, 167

Collinge, M. J., Sumi, T., & Fabrycky, D. 2006, ApJ, 651, 197

Hurta, Zs., Jurcsik, J., Szeidl, B., Sódor. Á. 2008, AJ, 135, 957

Jurcsik, J., Hurta, Zs., Sódor, Á., et al. 2009a, MNRAS accepted

Jurcsik, J., Sódor, Á., Hurta, Zs., et al. 2008, MNRAS, 391, 164

Jurcsik, J., Sódor, Á., Szeidl, B., et al. 2009b, MNRAS, 393, 1553

Kolenberg, K., Guggenberger, T., Medupe, T., et al. 2009, MNRAS, 396, 263

Moskalik, P., & Poretti, E. 2003, A&A, 398, 213

Stothers, R. 2006, ApJ, 652, 643

Comm. in Asteroseismology
Volume 160, September 2009
© Austrian Academy of Sciences

# Effects of resolution and helium abundance in A star surface convection simulations

F. Kupka[1,4], J. Ballot[2,4], and H.J. Muthsam[3]

[1] Observatoire de Paris, LESIA, CNRS UMR 8109, F-92195 Meudon, France
[2] Laboratoire d'Astrophysique de Toulouse-Tarbes, Université de Toulouse, CNRS,
14 avenue Edouard Belin, F-31400 Toulouse, France
[3] Faculty of Mathematics, Nordbergstr. 15,
A-1090 Vienna, Austria
[4] formerly at MPI for Astrophysics, Karl-Schwarzschild Str. 1,
D-85741 Garching, Germany

## Abstract

We present results from 2D radiation-hydrodynamical simulations of fully compressible convection for the surface layers of A-type stars with the ANTARES code. Spectroscopic indicators for photospheric convective velocity fields show a maximum of velocities near $T_{eff} \sim 8000$ K with the largest values measured for the subgroup of Am stars. Thus far, no prognostic model, neither theoretical nor numerical, is able to exactly reproduce the line profiles of sharp line A and Am stars in that temperature range. As the helium abundance of A stars is not known from observations, we have considered two extreme cases for our simulations: a solar helium abundance as an upper limit and zero helium abundance as a lower limit. The simulation for the helium free case is found to differ from the case with solar helium abundance by larger velocities, larger flow structures, and by a sign reversal of the flux of kinetic energy inside the hydrogen ionisation zone. Both simulations show extended shock fronts emerging from the optical surface, as well as mixing far below the region of partial ionisation of hydrogen, and vertical oscillations emerging after initial perturbations have been damped. We discuss problems related to the rapid radiative cooling at the surface of A-type stars such as resolution and efficient relaxation. This includes identifying the radiative cooling rate $Q_{rad}$, which poses the most severe time step limitation for the dynamical equation for the evolution of energy density, as a stiff term. It is thus a numerical rather than a physical constraint for the numerical simulation of these objects. The present work is considered as a step towards a systematic study of convection in A- to F-type stars, encouraged by the new data becoming available for these objects from both asteroseismological missions and from high resolution spectroscopy.

Accepted:    2009, September 22

## 1.  Introduction

Research on A-type stars has benefitted enormously from the introduction of high-resolution spectrographs with $R \gtrsim 100,000$ and from advancements in asteroseismology through telescope networks and satellite missions. They provide a new way of probing our theoretical understanding of these objects and address questions such as: how are convection and pulsation linked to each other, in a physical parameter region no longer completely dominated by the strong envelope convection found in cool stars such as our Sun? How does convective mixing in these stars affect diffusion processes? What physical mechanism lets chromospheres disappear in early A-type stars? How much does convection influence observed spectra and how does it interfere with determining stellar parameters from spectroscopy? This selection of physical questions demonstrates that A stars provide a unique physical laboratory (Landstreet, 2004) underlying the richness observed in their spectra (Adelman, 2004).

In the following, we report on the first results from a project on numerical simulations of A-type (and F-type) stars, motivated by the progress made by asteroseismological missions such as CoRoT and MOST and challenged by results from high-resolution spectroscopy. The simulations ultimately aim at improving our understanding of the dynamical processes of convection and its interaction with pulsation and, through mixing, on diffusion processes that shape the appearance of A-type stars. We first provide a background on how convection has been identified observationally in A-type stars. We then summarise the theoretical explanations given for the peculiar convective line broadening found in A stars and review previous attempts made to model the surface convection zones in these stars. In the subsequent section, we discuss the setup of the numerical simulations we have performed for mid A-type stars followed by a discussion of our results and our conclusions as well as an outlook on further work.

## 2.  Observational Background

The existence of convection in the surface layers of A stars was first predicted by Siedentopf (1933). Initially, observational evidence for such convection zones came from a non-zero microturbulence parameter $\xi$. Common values for $\xi$ were found in the range of $2 \ldots 4$ km s$^{-1}$ for both normal A and Am-type main sequence stars (see Landstreet 2007 for a review and Adelman 2004 for examples and further references). This readily exceeds the microturbulent velocities found for their cool star counterparts (see Gray 1988 for an extended discussion of $\xi$ in cool stars). To find direct evidence for convective velocity fields in the shapes of spectral lines of A stars is much more difficult. The average projected equatorial rotational velocities $v_e \sin i$ of normal A-type stars are found to be $> 120$ km s$^{-1}$ and even for Am stars values of less than $10 \ldots 15$ km s$^{-1}$ are the exception and not the rule (cf. the tables and references shown in Adelman 2004 and also the recent study of Royer et al. 2007). Hence, first evidence that the bisectors of spectral lines in early type (sharp line) stars do not resemble those observed for cool stars was found for late F-type supergiants (Gray & Toner, 1986) and the early F-type bright giant Canopus (Dravins & Lind, 1984; Dravins, 1987). Stars in this parameter range were shown to feature line profiles with curved bisectors caused by extended depressions in the blue wing of spectral lines. This results in an asymmetric line shape inverted in comparison with line profiles of cool supergiants (and those of our Sun and other cool main sequence stars). First evidence for this feature to be present in main sequence A stars was reported by Gray & Nagel (1989) for the case of HR 178 (HD 3883, an Am star).

This was followed by a study of Landstreet (1998) who found two Am stars, HR 4750 (HD 108642) and 32 Aqr (HD 209625), to show large asymmetries and a depressed blue wing in profiles of strong spectral lines. Weak lines were found to have much smaller profile asymmetries or even none at all. The same effects were found to be slightly weaker in the case of the hotter A-type star HR 3383 (HD 72660). A hot A-star with similar properties, HR 6470 A (HD 157486 A), was identified by Silaj et al. (2005). The current status of observational evidence of velocity fields caused by surface convection in main sequence A stars is analysed in Landstreet et al. (2009), who added another five stars to this sample. The newly identified objects include a mid A-type star with strong line profile asymmetries as previously found only for Am stars (Gray, 1989; Landstreet, 1998). Based on the collected data, which involved also reobserving objects during the same night, in consecutive nights, in different runs, and in different wavelength ranges, they concluded that the properties of the line profiles of these stars are robust. Furthermore, they concluded that the only explanation for the line profiles which is consistent with the observational data is the presence of photospheric velocity fields not related to pulsation thus confirming the suggestion by Landstreet (1998). These velocity fields are present in both normal A and in Am stars, though the latter contain the most extreme cases, in agreement with their higher values of $\xi$. It is interesting to note that chromospheric activity indicators were identified for A-type stars with

$T_{eff} \lesssim 8200$ K (Simon et al., 2002) which roughly coincides with the effective temperature of those stars for which the largest profile asymmetries and overall convective line broadening have been identified (Landstreet et al., 2009).

## 3.  Theoretical Explanations and Previous Simulations

The most straightforward explanation for the "inverted" or "reversed" line bisectors has already been given by Gray & Toner (1986) for the case of supergiants who suggested that these could be caused by hot columns of rising gas outshining the cold downflows. This was further developed by Gray (1989) and Dravins (1990) who noted that the hot upflows need to cover only a small fraction to reproduce such line profiles. This idea was also discussed for the case of main sequence A stars by Landstreet (1998).

Could this be recovered from non-local models of convection or radiation-hydrodynamical simulations? A first study of the coupled convection zones caused by ionisation of H I (and He I) as well of He II was presented in Latour et al. (1981), who used a modal expansion of an anelastic approximation to the hydrodynamical equations to describe convective motions in a mid A-type star. They noted the two convection zones to be coupled by overshooting. Due to the large He abundance (Y=0.354) assumed in this work it is difficult to compare it with more recent calculations (the large value of Y might partially explain why they found the convective flux to be larger in the He II ionisation zone, a property not recovered in later work). Xiong (1990) presented solutions obtained from his non-local model of convection for an F-type, an A-type, and two late B-type stars. Two separate convection zones were found coupled by overshooting for the mid A-type star with a convective flux being larger for the photospheric convection zone. A solar He abundance (Y=0.28) was assumed, as in the study of Kupka & Montgomery (2002), who used the fully non-local convection model of Canuto & Dubovikov (1998) and Canuto et al. (2001) to construct envelopes of A stars for an evolutionary sequence of a 2.1 $M_\odot$ star and for an effective temperature sequence at three different metallicities. Except for the hottest A stars the two convection zones were found to be coupled by overshooting with the photospheric zone dominating in mid and late A-type stars (in agreement with the numerical simulations by Freytag 1995, see below). They also investigated photospheric overshooting and found a positive kinetic energy flux for the observable layers of mid and late A-type stars. Such a distribution agrees with the model of Gray (1989); Dravins (1990); Landstreet (1998), although no synthetic spectra were computed at that point.

The first radiation-hydrodynamical (RHD) simulations of a mid A-type star in two spatial dimensions (2D) with $49 \times 42$ grid points were presented in Sofia & Chan (1984) (radial resp. vertical coordinate listed first here and in the following). However, an impenetrable upper boundary condition had to be placed at optical depths $\tau$ of 0.22 to 2/3 in those simulations. This put the boundary layer right into the zone of partial ionisation of H I with its accompanying density inversion and pushed vertical velocities to zero in that layer. Consequently, the lower convection zone due to ionisation of He II was found to dominate even at a solar He abundance and no predictions of photospheric velocities could be made. The stably stratified layers in-between those two zones were found to be fully mixed. In Gigas (1989) a 2D RHD simulation for the parameters of Vega was presented. The model had $30 \times 36$ points, but extended to much lower optical depths of $\log \tau \sim -2.4$. No He ionisation was included in the equation of state and simulation time had to be limited to one hour of stellar time. Vertical oscillations were found in the photosphere instead of a pattern of up- and downflows. Interestingly, line profiles calculated using that simulation were found to be bent bluewards. However, a comparison with Landstreet (1998) reveals that although the resulting bisector looks similar to those found for strong lines in HD 72660, a strong asymmetry is also found for weak lines contrary to observations, as corroborated by the analysis of Landstreet et al. (2009), who included another four stars in that $T_{\mathrm{eff}}$ range of 8900 K to 10000 K in their analysis. Given the short simulation time and the reported periods of 5 to 15 min the observed oscillations could also have been transient. This is also implied by results of the first extended study of A stars by means of 2D RHD simulations by Freytag (1995) (see also Freytag et al. 1996), who found oscillations to be damped even for a $T_{\mathrm{eff}} \sim 9000\ K$ and a $\log(g) = 3.9$, i.e., at the blue edge of the $\delta$ Sct instability strip, while oscillations were clearly excited in an even cooler model. Most of the simulation in Freytag (1995) were carried out for a typical resolution of $65 \times 62$ grid points, with a few simulations doubling horizontal resolution, and one case reaching $95 \times 182$ points for a short run. A solar chemical composition was assumed throughout and partial ionisation was accounted for as well. The case of non-grey radiative transfer was considered for one model and found to have little influence on the model structure and the flow itself. For models with $T_{\mathrm{eff}} \lesssim 9000\ K$ the convective (enthalpy) flux was always found to be larger in the upper convection zone, which is caused by the ionisation of hydrogen (H I), than in the lower lying one, which originates from the double ionisation of helium (He II). In that case the two convection zones were also always found to be connected by overshooting. A characteristic property of the simulations were nevertheless rather high velocities in the lower convection zone with vertical root mean square ($v_{\mathrm{rms}}$) maxima often around 1 km s$^{-1}$. These maxima are related to strong vortices found in the lower convection zone. No synthetic spectra were computed from those simulations.

A surprising change to these developments came finally with the first RHD simulations in 3D for two mid A-type main sequence stars by Freytag & Steffen (2004) (with a $T_{eff}$ of 8000 K and 8500 K for a $\log(g)$ of 4.4). The grey approximation and a solar chemical composition were assumed. 90 to 110 grid points were used along the vertical direction and 180 points along the horizontal ones. Similar to the previous 2D RHD simulations and solar granulation simulations they followed a box-in-a-star approach in which a limited volume located near the surface of the star was considered for the calculations. Contrary to most of the preceding 2D simulations the extent of the simulation boxes considered by Freytag & Steffen (2004) were chosen such as to contain at least several up- and downflow structures along any horizontal direction, which explains the larger number of grid points. Nevertheless, their simulations revealed the usual granulation pattern, with some subtle differences. In the end this lead to line profiles very much looking like solar ones in contradiction to the observations by Landstreet (1998). The calculated line profiles were further analysed by Steffen et al. (2006). In a follow-up work a third 3D RHD simulation (for a $T_{eff}$ of 8000 K and a $\log(g)$ of 4.0) was presented by Kochukhov et al. (2007) using again the $CO^5BOLD$ code as in Freytag & Steffen (2004) and Steffen et al. (2006), but this time with a slightly higher resolution ($170 \times 220^2$). In addition, part of the run was repeated using non-grey radiative transfer (based on binned opacities). Due to the latter the match for weak lines of stars such as HD 108642 (with actual stellar abundances used for the spectrum synthesis) improved. The same held for the cores of strong lines, but the wings of strong lines remained clearly discrepant in shape, width, and bisector curvature. Only one more 3D RHD simulation of a cool A-star ($T_{eff} = 7300$ K, $\log(g) = 4.3$, solar metallicity, non-grey) with $82 \times 100^2$ grid points was presented in more detail by Trampedach (2004), though without computations of stellar spectra. However, in that simulation the H I and He II zones are still connected to a single layer in a configuration described to be highly unstable. Further work in the literature has dealt with simulating interacting convection zones assuming idealised microphysics, but these results cannot be directly compared to stellar observations without invoking additional assumptions on the surface radiative cooling and the effects of chemical composition.

The discrepancy found by Freytag & Steffen (2004) is surprising, since RHD simulations using the same code had been used to compute realistic solar spectra that match observed line profiles (see Steffen 2007 and references therein). Obviously, the simulation results are in contradiction with the model of hot and narrow, rapidly rising columns of gas. The observations discussed by Landstreet et al. (2009) indicate that chemical peculiarity (Freytag & Steffen, 2004) can play only a limited role, as the line profile anomaly is found in both Am and normal A stars. This still cannot exclude He abundance to play a role (Kupka,

2005), at least at the level of explaining the differences between stars showing the most asymmetric and broadened profiles (of Am-type) and their counterparts among normal A stars. Effects of rotation might have to be considered, too (suggested by R. Arlt, see the conference discussion in Freytag & Steffen 2004; some consequences were analysed in Kupka 2005). This suggestion gains further importance by the frequent binary nature of the observed sharp line stars. However, the line profiles in Landstreet (1998) and Landstreet et al. (2009) do not resemble at all the flat-bottomed line profiles of rapid rotators seen pole-on such as Vega (Gulliver et al., 1994; Hill et al., 2004). Since there is no indication from the data on the sample of Landstreet et al. (2009) that the binaries are very close or interacting, nor that there is any dependence of the profiles on the rotation period other than the ability of detecting them, rotation appears to be a less likely candidate to resolve this discrepancy. Rather, numerical resolution (Freytag & Steffen, 2004) or the entire scheme used for the radiative transfer (Landstreet, priv. comm. 2004) may play a much more important role (see also Kupka 2005).

## 4.  Numerical Simulations of A-type Stars with ANTARES

To investigate the cause of these discrepancies between observation and numerical simulations we have decided to perform RHD simulations of A-type stars with the ANTARES code (Muthsam et al. 2007, Muthsam et al. 2009). This work is part of a more extended research project on convection in A- and F-type stars intended to study convection-pulsation interaction and the properties of convection under physical conditions that lead to very different efficiencies for the transport of heat and for the mixing of fluid, respectively.

As a first step we decided to have a closer look at the effects of helium abundance and of resolution, since both have not been investigated yet in sufficient detail. In addition, we also wanted to clarify the nature of the extremely restrictive time steps due to the short radiative cooling time scale $t_{rad}$ (Freytag & Steffen, 2004; Steffen, 2007) (see Eq. (7)). As ANTARES currently solves the fully compressible equations of hydrodynamics coupled to the stationary radiative transfer equation with a fully explicit time integration method, the ensuing time step constraints due to $t_{rad}$ are severe and the resulting computational costs very high (10 to 100 times higher than for a solar granulation simulation of comparable resolution according to Freytag & Steffen 2004). We have thus begun our study with RHD simulations in 2D to single out the most promising cases for 3D simulations and possibly suggest improvements to the time integration to reduce the computational costs of 3D simulations.

With ANTARES we solve the RHD equations for fully compressible flows,

$$\frac{\partial \rho}{\partial t} + \nabla \cdot [\rho \mathbf{u}] = 0, \tag{1}$$

$$\frac{\partial \rho \mathbf{u}}{\partial t} + \nabla \cdot [\rho \mathbf{u}\mathbf{u} + p\underline{I}] = \rho \mathbf{g} + \nabla \cdot \underline{\tau}, \tag{2}$$

$$\frac{\partial e}{\partial t} + \nabla \cdot [\mathbf{u}(e + p)] = \rho(\mathbf{g} \cdot \mathbf{u}) + \nabla \cdot (\mathbf{u} \cdot \underline{\tau}) + Q_{rad}, \tag{3}$$

which are the continuity, Navier-Stokes, and total energy equation describing the conservation of mass density $\rho$, momentum density $\rho\mathbf{u}$, and total energy density $e$. The last is the sum of internal ($e_{int}$) and kinetic energy density and $\mathbf{u}$ is the flow velocity. An equation of state (EOS) relates $e_{int}$ and $\rho$ to the temperature $T$ and the gas pressure $p$ (or rather, the sum of gas pressure and isotropic radiation pressure $p_{rad}$, the meaning of $p$ as used in the following). The EOS is taken from the most up-to-date OPAL tables (Rogers et al., 1996) and used also to compute all other required thermodynamic derivatives. We note that $\underline{I}$ is simply the identity matrix, $\underline{\tau}$ is the viscous stress tensor, and $t$ and $\mathbf{x}$ denote time and space. Neglecting bulk viscosity the viscous stress tensor components classically read

$$\tau_{ij} = \eta \left[ \partial_i u_j + \partial_j u_i - \frac{2}{3} (\partial_k u_k) \delta_{ij} \right], \tag{4}$$

where $\delta_{ij}$ denotes the Kronecker symbol and $\eta$ is the dynamic viscosity. In the upper part of stellar atmospheres, the radiative viscosity $\eta_{rad}$ dominates the molecular one (see, for instance, Chapman 1954). Hence, we have considered

$$\eta = \eta_{rad} = \frac{4\sigma T^4}{c^2 \kappa_{ross} \rho}, \tag{5}$$

where $c$, $\sigma$ and $\kappa_{ross}$ are the vacuum speed of light, Stefan-Boltzmann constant, and Rosseland mean opacity (Kippenhahn & Weigert, 1994) (this expression overestimates $\eta_{rad}$ for optically thin layers). Interestingly enough, the inclusion of $\eta_{rad}$ is sufficient to stabilize the simulations both in the solar case (where it remains small even in the photosphere) and for A stars (where it becomes large near the top of the simulations). This shows both the numerical stability and the very low numerical viscosity of ANTARES (see also Muthsam et al., 2009).

## 4.1.   Numerical methods

We give a short description of the numerical method used for the hydrodynamic equations and refer the reader to Obertscheider (2007) and Muthsam et al. (2009) for more details. – For the hyperbolic part of the equations (i.e., Euler equations) we use a fifth order WENO (weighted essentially non-oscillatory) method (Liu et al., 1994). Generally, the schemes which we discuss here are conservative numerical schemes. They use the principle of upwind biased interpolation. In order to achieve high-order accuracy and, at the same time, to capture discontinuities and steep gradients (essentially) without artificial oscillations a special method is applied to interpolate the necessary quantities, in particular fluxes, from the cell center to the cell boundary. The multidimensional problem is first split into one-dimensional problems (without thereby limiting spatial order of accuracy to two). The type of one-dimensional upwind biased interpolations in the direction of the coordinate axes constitute the core of the method. For a well-defined upwind direction the dependent variables are transformed into the characteristic system. The characteristic variables typically move with the velocities $u$ and $u \pm c$. Here, $u$ denotes the speed of the gas in the coordinate direction at hand and $c$ is the sound speed. Thus, for each characteristic variable an upwind direction is defined at each point depending on the sign of the relevant speed. − By increasing the number of interpolation nodes used to create a stencil the interpolation procedure is then built up hierarchically as follows. Initially, simple linear interpolation is performed using two grid points chosen such that the stencil points into the upwind direction. For quadratic interpolation the next gridpoints to the left and to the right of the interval and the *two* corresponding interpolation parabolas are considered. The smoother one among the two interpolation parabolas and the grid point belonging to it are adopted, the other grid point is discarded. Setting out with the new stencil containing three gridpoints a fourth grid point is now used and the corresponding interpolating cubic polynomial is determined in a similar way. This process is continued to any desired order. − In the *weighted* variant of these schemes (WENO), as used in this work, no grid point is discarded. Rather, one considers the set of polynomials corresponding to *all* stencils of the desired final length the unweighted method might encounter. For actual interpolation, a convex combination of these polynomials is used. The *weights* of the combination are determined in such a way that essentially maximum order of approximation is reached in the smooth part of the flow. If on the other hand there should be a discontinuity or a strong gradient, only those stencils pointing away from the discontinuity actually contribute. In this way, both highest achievable accuracy in the smooth part and stability plus still high order of accuracy of the scheme near discontinuities are attained.

In well-resolved calculations the method does not require any artificial viscosity to be added for stability reasons provided the contributions by radiative viscosity as given by Eq. (4) and (5) have been accounted for. We have also, in these calculations, not added any subgrid modelling, say of the Smagorinsky type. We feel justified to do so because in connection with our calculations of solar convection (granulation) as described in Muthsam et al. (2007) and Muthsam et al. (2009), we have repeatedly modelled the same physical situation with various grid spacings. The models on the coarse grid nicely resembled the better resolved ones and no need for a detailed subgrid modelling was obvious for this type of simulation. In contrast to subgrid models which try to model physical processes at small scales, hyperviscosities basically are designed just to prevent the simulations from blowing up. If applied anyway, they markedly degrade the quality of the numerical solution, see Muthsam et al. (2009), and have therefore not been applied in the simulations presented here either.

We now turn to the radiative contribution. In optically thick layers the radiative (heating and) cooling rate $Q_{rad} = \nabla \cdot F_{rad}$ can be computed from the diffusion approximation $F_{rad} = -K_{rad} \nabla T$, where $K_{rad}$ is the radiative conductivity obtained from equation of state quantities and the Rosseland opacity $\kappa_{ross}$. Numerically, $Q_{rad}$ is determined by central differences of high order. For the Rosseland opacities, we use tabulated OPAL opacities (Iglesias & Rogers, 1996) extended at low temperatures with tables by Ferguson et al. (2005), both computed for the solar mixture of Grevesse & Noels (1993). For the photospheric layers a radiative transfer equation must be solved for accurate computation of $Q_{rad}$. This computation is extended into deeper layers to allow a smooth transition to the diffusion approximation. Details on this procedure as used in ANTARES are given in Obertscheider (2007) and in Muthsam et al. (2009). We note that even in the grey approximation the results of this approach are different from the plain diffusion approximation. In ANTARES the stationary limit of the radiative transfer (intensity) equation is assumed (as in Freytag 1995, Freytag & Steffen 2004, Kochukhov et al. 2007) and the resulting $Q_{rad}$ is computed from an integration of the intensity obtained for a finite set of rays from this procedure. The grey case differs from the non-grey one by just integrating along each ray for a single frequency band rather than for multiple ones, but contrary to the diffusion approximation this accounts for inhomogeneity of cooling and heating in a convective stellar photosphere. We also point out here that the assumption of stationarity (instantaneous radiative transfer) leaves $\rho$, $\rho \mathbf{u}$, and $e$ as the dynamical variables of the system which have to be followed by the time integration during the simulation.

## 4.2. Setting up the simulations

The simulations discussed below are box-in-a-star simulations for a limited volume ranging from the upper photosphere to the stably stratified layers around $T \sim 60,000$ K to $70,000$ K, thus containing the surface convection zones inside the box. Since this region is small with respect to the stellar radius, a Cartesian (box) geometry is taken for the simulation box and the gravity $\mathbf{g}$ is a constant, downwards pointing vector equalling the surface gravity $g$. Horizontal boundary conditions are hence periodic and the geometry is plane parallel. The horizontal domain is chosen wide enough to fit several up- and downflow structures into the box at photospheric depth. The vertical boundary conditions are assumed to be impenetrable and stress-free (free-slip) with a constant energy flux assumed at the bottom of the simulation zone (and directly related to the $T_{eff}$ defined for the simulation). The lower boundary is placed sufficiently far below the convection zones to have only limited influence (at most through reflecting waves). The upper boundary condition is placed at $\log \tau_{ross} \sim -5$ for similar reasons. The appropriate domain sizes are estimated from scaling the linear dimensions of solar granulation simulations by the pressure scale height $l = H_p = P/(\rho g) \sim T/(\mu g)$, which when comparing to the Sun implies

$$ l/l_\odot = T_{eff}\mu_\odot g_\odot / (T_{eff,\odot}\mu g) \tag{6} $$

for a star characterised by $(T_{eff}, \mu, g)$, where $\mu$ is the mean molecular weight.

To perform simulations which are close to an observational case we have chosen $T_{eff} = 8200$ K and a $\log(g) = 4.0$ as basic physical parameters, as they are representative of HD 108642 (Kupka et al. 2004; note that Landstreet 1998 and Landstreet et al. 2009 suggested a $T_{eff}$ 100 K lower and a $\log(g)$ 0.1 dex higher than those values, but such differences are well within the observational uncertainties). Since diffusion has been suggested to substantially change the helium content of the upper envelope during the main sequence life time of A-type stars (Vauclair et al., 1974; Richard et al., 2001; Michaud, 2004), we consider two extreme cases of helium abundance: a solar one and no helium at all. A solar metallicity mass fraction of $Z = 0.02$ was taken in both cases (this value is a compromise, because the underabundance of light elements relative to their solar values lowers the absolute value of $Z$ while the opacity resulting from the different chemical composition contributing to $Z$ is increased in the photosphere of Am-type stars; see Kupka et al. 2004 for further details on the abundances of HD 108642). The hydrogen mass fraction $X$ is scaled up from 0.7 to 0.98 in the case where helium is absent ($Y = 0$).

For the simulations presented here we have chosen a coarse grid with a single refinement zone embedded inside the domain covered by the coarse grid. Since the simulations are in 2D, the horizontal boundary conditions of the coarse grid are periodic and because the fine grid is used to improve resolution near the surface, it has the same horizontal extent as the coarse one. Hence, it has periodic boundary conditions as well. Vertically, the fine grid is located in the interior of the coarse grid which provides the vertical boundary conditions for the fine grid. The coarse grid itself has the closed vertical boundary conditions already described above. For the case with solar helium abundance, the coarse grid has 150 × 200 points which span a domain of 32.2 × 40.3 Mm² and yield a resolution of $216 \times 201$ km². The fine grid has $205 \times 200$ points and is located vertically between about 1.1 Mm and 12.1 Mm as measured from the top of the simulation box. This provides a resolution of $54 \times 201$ km², i.e., a vertical refinement by a factor of 4. For the plots shown below the vertical axis is shifted by $-3.8$ Mm to have the downwards pointing vertical axis be labelled with zero where the horizontally and temporally averaged temperature equals $T_{\text{eff}}$. For the simulation without helium, a coarse grid of $130 \times 210$ points is used spanning a domain of $26.1 \times 41.7$ Mm² and thus providing a mesh size of $202 \times 199$ km². The fine grid is located vertically between about 1.8 Mm and 13.1 Mm as measured from the top and has $225 \times 210$ grid points. Hence, the resolution on the fine grid is $51 \times 199$ km², which again implies a vertical refinement by a factor of 4. For this simulation the vertical axis is shifted in the plots shown below by $-4.6$ Mm. As a result, in plots showing both simulations the vertical axis value of zero indicates the average location where $T = T_{\text{eff}}$, i.e., the optical surface, for both cases. For the radiative transfer, we have used 12 rays and a single bin for Rosseland mean opacities, i.e., the grey approximation.

With these numerical values we can also estimate how close our numerical simulations are to a regime which permits the development of turbulence through shear between the up- and downflows on the resolved length scales. Following the discussion in Kupka (2009) we compute an effective Reynolds number $\text{Re}_{\text{eff}} = UL/\nu_{\text{eff}}$ which compares the inertial forces for velocities $U$ and sizes $L$ at the size of up- and downflow structures near the surface with the effective viscosity $\nu_{\text{eff}}$ of the simulation achievable for a finite grid size $h$. For a 3D simulation this can be estimated from the relation $\text{Re}_{\text{eff}} \approx (L/h)^{4/3}$ (see Kupka 2009 and its discussion of a detailed derivation in Lesieur 1997). From (6) and a solar surface granule size we estimate $L \approx 4666$ km and $\approx 9333$ km for the solar helium and the helium free composition, respectively, assuming the stellar parameters given above. A 3D simulation with $h = 200$ km would then imply fairly moderate values of $\text{Re}_{\text{eff}}$ close to 70 and 170, respectively, and a resolution $h = 50$ km would improve this to about 380 and 1070 for both cases, thus making the simulations moderately turbulent due to shear on resolved scales. However, since our simulations have been made in 2D, there is a smaller number of internal degrees of freedom and the appropriate scaling law is actually $\text{Re}_{\text{eff}} \approx (L/h)^2$ (see again Kupka 2009 and its discussion of results derived in Lesieur 1997). Consequently, the simulations rather have $\text{Re}_{\text{eff}}$ of 470 and 2180 for the present horizontal resolution which puts them into the transition regime of a fully turbulent flow. If the vertical resolution of 50 km were also applied horizontally, the simulations would be fully turbulent, with $\text{Re}_{\text{eff}}$ values of 7500 and 34800. In the same way we can define an effective Prandtl number $\text{Pr}_{\text{eff}} = \nu_{\text{eff}}/\chi$, where $\chi$ is the (radiative) thermometric conductivity. With $\mathrm{Pr} = \nu/\chi$ as the Prandtl number and $\mathrm{Re} = UL/\nu$ as the Reynolds number this can be rewritten as $\mathrm{Pr_{eff}} = \mathrm{Pr}\,\mathrm{Re}/\mathrm{Re_{eff}}$. Taking $\mathrm{Pr} \sim 10^{-8}$ as an upper limit and $\mathrm{Re} \sim 10^{10}$ as a lower one, we have that $\mathrm{Pr_{eff}} \lesssim 100/\mathrm{Re_{eff}}$. A pessimist's estimate would hence expect $\mathrm{Pr_{eff}}$ to be around 1 for the 3D simulation and $h = 200$ km and in the range of 0.1 to 0.25 at 50 km. For the 2D simulation we already are in the range of 0.05 to 0.2 for our present simulations and $h = 50$ km for both horizontal and vertical directions would push this below about 0.01. We thus expect our simulations to be already in the regime of physical interest at the present resolution, i.e. that of a turbulent low Prandtl number flow. For the 3D case it would be more important to reach a resolution of 50 km also in the horizontal direction. At this point we would like to mention that for A stars the photospheric radiative viscosity, Eq. (4)–(5), is large enough to lower the physical Reynolds number of the photospheric flow by several orders of magnitudes, well into a laminar to just mildly turbulent regime, although the underlying estimate is necessarily crude, since $\eta_{\mathrm{rad}}$ is computed from a diffusion model.

Apart from setting up the simulations and some test runs which were done on single and two CPU core configurations most of the simulation runs to evolve both the solar helium and the helium-free case were done on 8 CPU cores using the MPI capabilities of ANTARES. The code was slightly modified to allow grid refinement zones to cover the entire horizontal extent of the model, as the fine grid was previously used to have higher resolution only in a specific domain of interest (cf. Muthsam et al. 2007). These enhanced possibilities to use fine grids are now a standard feature of the code (Muthsam et al., 2009). Another improvement added to the code and motivated by simulations of A-type stars was its ability to work with arbitrary chemical compositions (and thus combinations of $X, Y$, and $Z$), since the opacity and equation of state tables used have some restrictions on the combinations of these parameters. The improved microphysics interface is now also a standard module of the code. After initializing the simulations, all production runs were performed on the POWER5 p575 of the RZG in Garching. The initial conditions for the simulations were taken from 1D models for HD 108642 calculated with the LLmodels code (Shulyak et al., 2004) using the convection model by Canuto & Mazzitelli (1991) and opacity distribution function tables by Kurucz (1993a) and Kurucz (1993b). For the 1D models $\xi = 4$ km s$^{-1}$ was assumed (Landstreet, 1998; Kupka et al., 2004) and $T_{\mathrm{eff}}$ and $\log(g)$ were the same as in the simulations (values taken from Kupka et al. 2004). The flow in the simulations is started by a lack of perfect hydrostatic equilibrium of the stratification introduced by the differences in the equation of state and opacities used, the change in grid spacing between the 1D model and the simulation, and the different numerical methods used in both codes. In addition, we add a small perturbation in the mass density field by creating a random distribution for the density perturbations in Fourier space. The values of this smooth noise function are added at each point to the average for the horizontal layer as obtained from the 1D model. The noise is equally distributed over all modes (wave numbers), but its magnitude is scaled as a function of depth to produce the largest density fluctuations where the 1D convection model predicts the largest temperature fluctuations. This avoids the introduction of long-lived perturbations in the radiative region near the bottom of the simulation box, which would cause unnecessarily long relaxation times (we return to this topic further below).

## 5. Results

In the following we discuss the main results of our 2D RHD simulations for both the case with solar helium abundance and for the case with zero helium abundance. Some results from experimental runs with different initial conditions and resolution but otherwise identical parameters are included as well.

### 5.1. Relaxation and oscillations

Both simulations were run for several sound-crossing times until convective motions had sufficiently developed, which is indicated by a growth in vorticity and in total kinetic energy. For the simulation with solar He abundance notable growth of convective motions occurred after about three sound-crossing times (time for a sound wave to cross the entire simulation box vertically, here $t_{sc} = 1371.6$ s). After $t/t_{sc} \gtrsim 5$ convective motions dominated the contributions to kinetic energy and experienced a phase of exponential growth. Fig. 1 shows the velocity of the centre of gravity of the entire simulation volume as a function of time. The oscillations triggered by the stratification being out of perfect hydrostatic equilibrium are prominent and are almost purely vertical since the horizontal component is totally negligible. There is only weak intrinsic damping. We attribute this to the quality of the advection scheme implemented in ANTARES. We removed these oscillations by introducing a damping term at $t/t_{sc} \sim 6.1$ for the vertical velocity component, as is routinely done so during relaxation of solar granulation simulations (Trampedach, 1997). Due to the short time scale chosen the oscillations are rapidly damped which allows us to turn off this artificial damping after $t/t_{sc} \sim 7.9$. At this point the vertical oscillations contribute $\lesssim 3\%$ to the total kinetic energy of the flow and shortly afterwards the total kinetic energy starts dropping for the first time after its rapid phase of growth and at least the upper convection zone itself is essentially relaxed. The snapshots in Figs. 5, 7, and 8 have hence been taken at $t/t_{sc} \gtrsim 8$. The equivalent holds for the averages in Fig. 3 and 4.

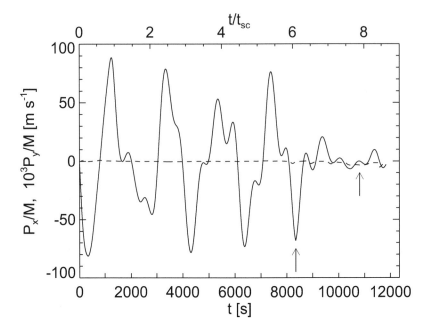

Figure 1: Vertical (solid line) and horizontal (dashes) velocities of the centre of gravity for the simulation with solar helium abundance as a function of time. We denote $P_x$ and $P_y$ the total vertical and horizontal momentum of the simulation and $M$ its total mass. The horizontal component is magnified by a factor of $10^3$. On the top x-axis, time is normalised by the sound-crossing time. Arrows indicate the beginning and the end of the artificial damping phase.

An alternative way of visualizing vertical oscillations is to look at the motion of the photosphere and define the velocity $v_{ph}$ as the horizontally averaged vertical motion of the layer for which $T = T_{eff}$ (the photospheric radius) at a given point in time. For both simulations this velocity is about an order of magnitude larger than that of the centre of gravity. Fig. 2 shows the time development of $v_{ph}$ for the simulation with zero helium abundance. Intrinsic damping is just slightly more efficient than in the previous case. Artificial damping of vertical motions is applied from $t/t_{sc} \sim 4.4$ to $t/t_{sc} \sim 13.9$ onwards (here, $t_{sc} = 1150.9$ s). As we have used a smaller damping rate for this second simulation the damping phase is longer and smoother than for the previous case. The development of convection also takes longer: strong convective motions set in only after $t/t_{sc} \sim 7$ and they dominate over the vertical oscillations in terms of kinetic energy after $t/t_{sc} \sim 9$. While the convective motions increase in strength, the contributions by oscillations drop until they reach a level of

$\lesssim 3\%$ at $t/t_{sc} \sim 14$. Once the flow can evolve undamped, some oscillatory motions seem to rise again. This is more obvious for $v_{ph}$ than for the centre of gravity, despite $v_{ph}$ is also much more influenced by events taking place at the surface such as the formation of shock fronts, which contribute to the non-sinusoidal shape of the function visible in Fig. 2. With kinetic energy increasing no longer exponentially the simulation slowly reaches equilibrium at $t/t_{sc} \sim 18$. The snapshot in Fig. 6 and the averages in Fig. 4 have been computed for $t/t_{sc} \gtrsim 17$.

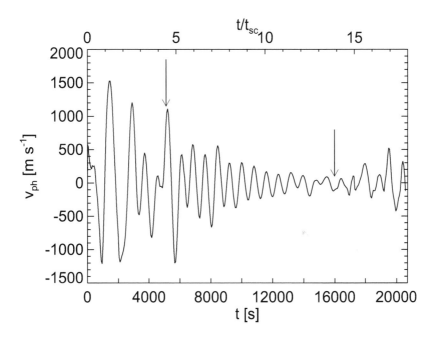

Figure 2: Temporal evolution of photospheric velocity for the simulation with zero helium abundance. On the top x-axis time is normalised by the sound-crossing time. Arrows indicate the beginning and the end of the artificial damping phase.

We notice that a proper choice of the initial perturbations is more important for simulations of surface convection in A stars than for solar granulation simulations. An earlier model for the case of solar helium abundance, which had been perturbed with an additional, sinusoidal velocity field instead of a random perturbation in density, showed strong resiliency to "forget" this pattern in the zone of He II ionisation even after several sound-crossing times. That simulation run was hence discarded. This behaviour is quite different from the solar case where the same kind of perturbation is rapidly removed by the convective flow, for instance, in the simulations of Obertscheider (2007).

## 5.2. Structure of the convection zone

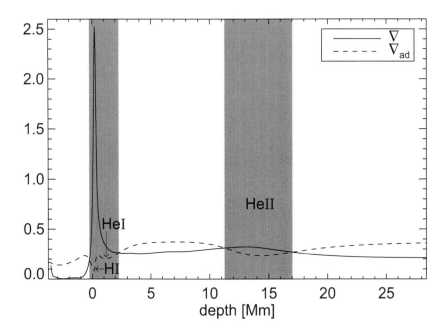

Figure 3: Temporal and horizontal average of actual ($\nabla$) and adiabatic ($\nabla_{ad}$) temperature gradient for the simulation with solar helium abundance. The regions of partial ionisation of H I, He I, and He II are indicated. Layers unstable according to the Schwarzschild criterion are indicated in grey.

In Fig. 3 we show the dimensionless temperature gradient $\nabla$ and the dimensionless adiabatic gradient $\nabla_{ad}$ for the simulation with solar helium abundance. The superadiabatic peak in the zone of partial hydrogen ionisation is about 3.5 times steeper than in a simulation of solar granulation (cf. Rosenthal et al. 1999 for the latter). The upper convection zone is about 2.52 Mm deep as obtained from the definition of the Schwarzschild criterion applied to the horizontally and time averaged gradients. The upper zone is mostly driven by the partial ionisation of H I. It is extended due to the partial ionisation of He I which takes place close enough to unite both regions into a single convection zone.

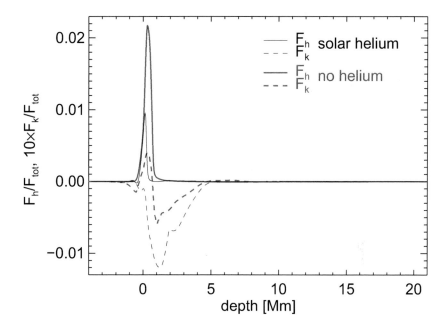

Figure 4: Convective (enthalpy) $F_h$ and kinetic energy flux $F_k$ in units of the total flux, shown as solid and dashed line, respectively, for the simulation with solar helium abundance (black thin) and without helium (grey thick). The kinetic energy flux is multiplied by a factor of 10 for better visibility.

With 5.7 Mm the second convection zone, which is caused by partial ionisation of He II, is much larger in vertical extent. This is mostly due to the increase of the local pressure scale height. The resulting mean structure is very similar to that one found in one-dimensional models of stellar atmospheres and stellar envelopes for that part of the HR diagram (cf. Vauclair et al. 1974, Latour et al. 1981, Landstreet 1998), because the high radiative losses of convection keep the stratification of the entire envelope in these stars close to that one of a purely radiative model (this has also been found by Freytag 1995, Freytag et al. 1996, Kupka & Montgomery 2002, Freytag & Steffen 2004).

By comparison, the simulation with zero helium abundance has just a single convection zone due to the partial ionisation of H I. Its extent into the photosphere up from where on average $T = T_{eff}$ is marginally smaller than in the previous case (0.23 Mm instead of 0.25 Mm). With a total extent of 1.56 Mm when following the same definition as used above, it is more shallow than its counterpart containing helium, in spite of a larger local scale height due to a

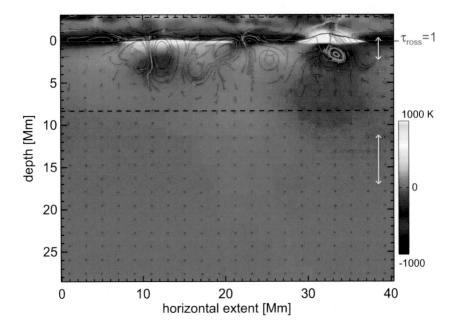

Figure 5: Snapshot of temperature fluctuations ($[T'] = K$) relative to their horizontal mean for the simulation with solar helium abundance. Extreme values have been chopped for easier visualization. The two black dashed lines indicate the region where grid refinement has been applied. The $\tau_{\mathrm{ross}} = 1$ 'isosurface' is denoted by a solid line. Streamlines indicate the direction and magnitude of the local flow field (the mean vertical velocity has been subtracted to improve visibility of the weakest vortices). The vertical extent of the convection zones according to the Schwarzschild criterion is indicated by white arrows.

lower $\mu$, Eq. (6). This is a direct result of the lack of He I ionisation extending the zone deeper inside (note the behaviour of $\nabla_{\mathrm{ad}}$ in Fig. 3). However, with a maximum of 1.6 the gradient $\nabla$ itself is flatter and the resulting superadiabatic gradient is only about 2.3 times steeper than in solar granulation simulations.

Differences are also visible in the flux distribution. Fig. 4 displays the enthalpy (convective) flux $F_{\mathrm{h}}$ and the flux of kinetic energy $F_{\mathrm{k}}$ for both simulations in units of the total (input) flux after subtracting the contribution due to the mean flow (i.e., due to the remaining vertical oscillations). That contribution is negligible for $F_{\mathrm{k}}$, but can contribute a relative flux of up to 0.5% to $F_{\mathrm{h}}$, if the averaging time is short, i.e., $t < t_{\mathrm{sc}}$. The higher enthalpy flux found in the simulation with no helium is hardly significant since it could be the signature either of a more relaxed simulation or of an effectively stronger convection due

to a lower mean molecular weight. The same could hold for the higher velocities found for the helium free case, though we consider them more likely to be
caused indeed by a lower $\mu$. Nevertheless, and interestingly enough, we note
that in the simulation with no helium $F_k$ has the opposite sign for the upper
half of the convection zone. This feature remains robust even when averaging
for $t/t_{sc} \sim 3$ and also when averaging over one third of that time scale for
different subintervals. This inversion in the sign of $F_k$, which does not appear
in simulations of solar convection, occurs in our simulation with the most shallow convective zone. In this case, with a really thin convective layer, heating
from below and cooling from above act at the same time within a pressure
scale height, while in a thicker zone, such as in the model with solar helium
abundance, heating and cooling are more disconnected (cf. Moeng & Rotunno
1990 for a discussion of heating from below and cooling from above in a me-
teorological context). We also note that at the chosen scale the fluxes in the
He II ionisation zone are not visible (cf. Steffen et al. 2006 and Steffen 2007).
These small fluxes require a more careful inspection of the simulation.

Indeed, if we look at Fig. 5 we see that the strongest vortices created by
the flow are located in the upper convection zone or just slightly underneath it
and thus are close to the region of the strongest convective driving. The large
downdraft at a horizontal coordinate of 23 Mm manages to penetrate just a
little bit into the lower convection zone, but at that time no strong vortices have
developed in that region. We recall our previous observation that if vortices are
seeded there as an initial condition, for instance by a sinusoidal perturbation,
they remain there for a long time. This is both due to the weak convective
driving (which cannot rapidly destroy a vortex or push it into a stably stratified
layer) and to the general lifetime of vortex patches in 2D (cf. Muthsam et al.
2007). Temperature fluctuations are found largest at the optical surface (the
extrema have been truncated in the plot for a better contrast and can be four
to five times larger while root mean square fluctuations are twice as large). The
optical surface is corrugated mostly around rapidly evolving upflow regions.

In Fig. 6 we see quite a similar overall pattern for the simulation without
helium. The horizontal scales are somewhat larger, as expected from Eq. (6).
Vortex patches are also strongest inside the convection zone itself, but a larger
number of them is found further below than in Fig. 5. We think that this is a
result of the longer time evolution of this simulation. Eventually, also for the
scenario shown in Fig. 5 downflows from above will form new vortex patches in
the lower convection zone or patches located in the stably stratified layer in between will reach it. Thus, if we run the simulation with solar helium abundance
significantly longer, we may eventually see an increase in $F_h$. However, this may
not be more realistic, as the vortex patches in 2D just happen to assemble near
the bottom of a simulation of a strongly stratified medium (cf. Muthsam et al.

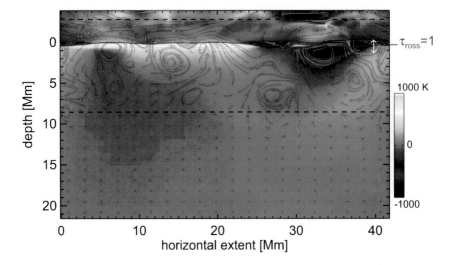

Figure 6: Snapshot of temperature fluctuations ($[T'] = K$) relative to their horizontal mean for the simulation without helium. Scaling of $T'$, domain boundaries of the grid refinement zone, the location of the optical surface defined at $\tau_{ross} = 1$, and the local velocity field (with mean vertical velocity subtracted) are indicated as in Fig. 5. The vertical extent of the convection zone according to the Schwarzschild criterion is indicated by a white arrow.

2007). Since 3D simulations started from 2D ones may inherit these properties for quite some (simulation) time, we think that future long-term simulations are the only way to accurately compute the enthalpy and kinetic energy fluxes in the lower convection zone of models for A stars with a helium content large enough to drive that zone.

## 5.3. Shock fronts and their development

As we can conclude from the small enthalpy fluxes found for both simulations shown in Fig. 4 convection is rather inefficient in transporting heat in mid A-type stars, a result also reported by Freytag (1995), Freytag et al. (1996), Kupka & Montgomery (2002), Freytag & Steffen (2004), Steffen et al. (2006), and Steffen (2007). Consequently, the mean temperature gradient stays close to the radiative one and hence, contrary to RHD simulations of solar granulation, a density inversion appears in the layer where H I is partially ionised (Freytag, 1995). This is a consequence of the lower mean molecular weight which in turn is caused by the doubling of the number of particles contributed by ionised hydrogen relative to neutral one. In the Sun efficient convection and thus a much flatter $\nabla$ is counterbalancing this effect of partial ionisation (cf. the simulations of M. Steffen and F.J. Robinson compared in Kupka 2009; the model by Canuto & Mazzitelli 1991 is an exception, since it predicts inefficient convection for the photospheric layers of the Sun which are located on top of a rapid transition to efficient, adiabatic convection dominating the bulk part of the convective envelope). Fig. 7 shows a snapshot for our simulation with solar helium abundance. The density inversion is prominent and the ensuing Rayleigh-Taylor instability enhances the convective instability caused by high opacity and a low $\nabla_{ad}$ (see Fig. 3 for the latter).

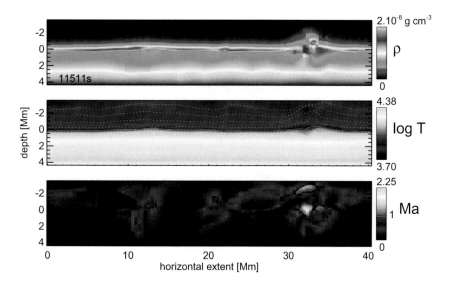

Figure 7: Mass density $\rho$, logarithm of temperature $T$ ($[T] = K$), and local Mach number in the simulation with solar helium abundance prior to the formation of a large shock front. The isoline for which $\tau_{ross} = 1$ is denoted in the middle panel by a thick solid line. Isolines for $\log \tau_{ross} = -4, -3, -2, -1, +1$ are denoted as dotted lines. Note the density inversion near the optical surface in the top panel. Regions of supersonic flow are surrounded by a bright solid isoline where $Ma = 1$.

Note that the inversion layer is interrupted by a complex feature at a horizontal coordinate of $\sim 33$ Mm. In this region the optical surface is more corrugated (solid line in the temperature plot in the middle panel) and supersonic velocities are found there, too (bottom panel). This feature is the result of two counterrotating vortices, which form at some time $t/t_{sc} \gtrsim 7.5$. At $t/t_{sc} \sim 8$ they have created an extended region just underneath the photosphere with a

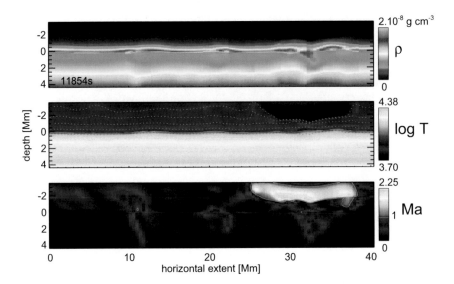

Figure 8: Mass density $\rho$, logarithm of temperature $T$ ($[T] = K$), and local Mach number in the simulation with solar helium abundance 343 s ($0.25\, t_{sc}$) after the time step displayed in Fig. 7. Note the large shock front in the photosphere spreading more than 1/3 of this layer within the simulation box. It is characterised by a large jump in temperature and a relative increase in the optical depth by more than a factor of 10. Isolines have the same meaning as in Fig. 7.

temperature lower than its horizontal average. The low temperature fluid is supplied by the strong downdraft between the vortex pair which reaches supersonic velocities in its inflow area within the photosphere. At $t/t_{sc} \sim 8.2$ hot gas from the neighbouring upflows has mostly covered the cold downflow region leaving a dense spot of cold material in its centre. As a result supersonic flow appears both in the two photospheric inflow areas and in the downdraft itself. From now on the structure begins to collapse emitting a series of shock fronts. At $t/t_{sc} \sim 8.4$, which is shown in Fig. 7, there are no more supersonic inflow regions. Rather, supersonic velocities are found in a small front in the upper photosphere (bottom panel), visible also in $T$ (band structure trailed by material of lower $T$ underneath) and in $T'$ in Fig. 5, which shows the state of the simulation 68 s before that one illustrated in Fig. 7. The streamlines in Fig. 5 also demonstrate that the structure is no longer dominated by a pair of roughly equally strong vortices. Supersonic flow is also found just underneath the patch of dense gas extending into the photosphere (Fig. 7).

Once this gas patch has bumped onto the layers just below it, a large shock front is created which horizontally extends over one third of the simulation box and also spreads over almost one third of the photosphere contained in the simulation volume. This can be seen in Fig. 8, at $\sim 0.25\, t_{sc}$ after the state shown in Fig. 7. The snapshot shows the moment when the remainders of the previous front, which is shown still moving upwards in Fig. 7 and is reflected by the upper boundary condition shortly afterwards, are absorbed by the new front. This explains the extended high Ma number region preceding the new front (we recall here that a shock front is a (near) discontinuity in the hydrodynamic variables, which does not necessarily have to move at supersonic velocities). As the upwards moving material has much higher density, it immediately absorbs the downwards moving one with very little changes to its physical state. Note that the shock front is optically thin, although with a steep gradient in optical depth (isolines of $\log \tau_{ross} = -3$ and $-4$ nearly merging with each other). At the same time the anomaly in the density stratification has nearly disappeared. Comparing with Fig. 7 we can see that the density in this region has dropped dramatically (in fact to average horizontal values) leaving a low temperature region which is reheated by the upwards moving shock front. We note that similar events also appear in the simulation run without helium and smaller events appear more frequently than the extreme case illustrated by Fig. 7 and Fig. 8. These violent events have no counterpart in simulations of solar convection performed with the ANTARES code (Muthsam et al., 2007). Even though they are intermittent in nature, they are common enough (cf. the front visible in Fig. 6 in the upper right corner of the simulation volume) to be statistically important. They may even contribute to the shape of line profiles and provide an efficient source of heating up a chromosphere (cf. Simon et al. 2002). Landstreet et al. (2009) estimate that at least locally velocities may have to be as high as $10...12$ km s$^{-1}$, which corresponds to Mach numbers of about 1.5, to reproduce the most extreme cases of line profiles in their sample of sharp line A and Am main sequence stars. This is easily within the range of the simulations presented here. We note that Shibahashi et al. (2008) have suggested shock wave trains to explain peculiarities in the line profile variations of roAp stars of similar $T_{eff}$ and $\log(g)$ as the A and Am stars which are the subject of our study. However, in their model the shock waves result from global stellar oscillations rather than being the by-product of a convective flow near the stellar surface.

## 5.4.   Resolution and computation of the radiative cooling rate $Q_{rad}$

Already for the case of solar granulation the most difficult quantity to resolve in equations (1)–(3) is the radiative heating and cooling rate $Q_{rad}$ (Nordlund & Stein, 1991). It directly influences the large, energy carrying scales by driving the cooling (and reheating) of gas in the photosphere. Quantities directly related to

$Q_{rad}$ such as the radiative flux $F_{rad}$ and the superadiabatic temperature gradient $\nabla - \nabla_{ad}$ or the profiles of $T$ and $p$ are readily resolved at vertical grid spacings of $\sim 12...25$ km, but to obtain an equally smoothly resolved $Q_{rad}$ requires a 10 times higher resolution (Nordlund & Stein, 1991), even though $T$, $p$, and $F_{rad}$ appear reasonably well converged at a resolution of $\sim 12...25$ km. Apparently, at that resolution the simulation runs benefit from some averaging effects. We also recall that in regions where the diffusion approximation holds, we have $Q_{rad} = -\mathrm{div}(K_{rad}\nabla T)$ which involves a second derivative of $T$. This explains why resolution requirements for $Q_{rad}$ are higher than for mean structure quantities ($p$, $T$, ...) and their first derivatives ($F_{rad}$, $\nabla$, ...).

How well do we resolve $Q_{rad}$ in A-type stars? Analysing our simulations for the coarse grid, which at 200 km resolution is equivalent to a solar granulation simulation with a resolution of $\sim 50$ km, the answer is: *at that grid spacing not at all*. Near the surface $Q_{rad}$ often has a triple-peak structure covered by about 15 grid points. At the same time on the fine grid there is only a double peak-structure ranging just 5 such grid points, i.e. 20 points on the fine grid, whereas the maxima of these peaks have less than $1/3$ of the size found on the coarse grid. Physically, the double peak-structure originates from the radiative cooling layer at the bottom of the photosphere (which is also present in RHD simulations of solar surface convection) and a heating layer just underneath it. The latter is created by the partial ionisation of H I and the inefficiency of convective transport in the superadiabatic layer coinciding with the entire surface convection zone (see Fig. 3). Once the simulation contains a fully developed surface convection zone, the detailed structure of the peak becomes more complex, as heating and cooling layers are no longer confined to a thin zone (cf. Fig. 5 to Fig. 8). Since the coarse grid solution is discarded by ANTARES at each time step for a domain where a fine grid solution is available, the wrong solution on the coarse grid has no impact on the development of the fine grid. But it nevertheless implies a warning concerning simulations of mid A-type stars with lower effective resolution for $Q_{rad}$ than the one presented by our fine grid solutions. We consider even our current simulations on the borderline of resolving the overall structure and magnitude of $Q_{rad}$, because each of the very sharp peaks is represented by only about 5 points. By comparison, at a vertical resolution of 16 km (corresponding to about 60 km for our present case) a solar granulation simulation would have the main (negative) peak covered by 15 points which resolve $Q_{rad}$ smoothly everywhere except near its extremum (Obertscheider, 2007). Thus, A stars require a 3 to 4 times higher effective resolution at the stellar photosphere to properly resolve $Q_{rad}$. This is also in agreement with the fact that $\nabla$ is steeper than in the Sun by just about that factor (Fig. 3).

There is a second problem associated with $Q_{rad}$. The radiative time scales in the photospheres of A stars are much shorter than in the Sun (Freytag, 1995; Freytag & Steffen, 2004; Steffen, 2007), since the opacity at their surface is lower while temperatures are higher. For optically thick layers the lower densities near their surface play a role, too. Hence, the time scale for relaxing a temperature perturbation of arbitrary optical thickness by radiation (Spiegel, 1957),

$$t_{rad} = \frac{c_v}{16\kappa\sigma T^3} \left(1 - \frac{\kappa\rho}{k} \operatorname{arccot} \frac{\kappa\rho}{k}\right)^{-1}, \qquad (7)$$

becomes smaller than the hydrodynamical time scales in a simulation. This includes the time scale of sound waves travelling a grid distance $h$, i.e., $t_{sound} = h/c_s$. In Eq. (7), $c_v$ is the specific heat at constant volume, $\kappa$ is the opacity, $\sigma$ is the Stefan-Boltzmann constant, and a perturbation of size $l$ with $k = 2\pi/l$ is assumed. For the optically thick case $t_{rad}$ converges to the time scale of radiative diffusion, $t_{diff} = 3(\kappa\rho/k)^2 c_v/(16\kappa\sigma T^3) \sim l^2/\chi$, but remains larger than zero for the optically thin case defined by $\kappa\rho/k \ll 1$ (using the specific heat at constant pressure, $c_p$, in the definitions does not change the argument). Comparing the time scales $t_{sound}$ and $t_{rad}$ for the case $h = l$ reveals the shortest radiative relaxation time scales in the problem. It is important to note that although for an A-type star the sound speed $c_s$ is larger than for the Sun in the layer where $T = T_{eff}$, $t_{rad}$ becomes even smaller. For the Sun at usual resolutions of $20 \ldots 30$ km for granulation simulations (Kupka, 2009) $t_{sound} \lesssim t_{rad}$, while for an A-type star with an equivalent resolution $t_{rad} \sim 0.01 \ldots 0.1\, t_{sound}$ for layers around the optical surface (Freytag, 1995; Freytag & Steffen, 2004).

As a consequence, RHD simulations of mid A-type stars, which use a purely explicit time integration method, are limited to time steps $\Delta t \leq t_{rad}$. This was noticed in Freytag & Steffen (2004) and also in Freytag (1995). Since ANTARES currently uses a purely explicit time integration scheme for Eqs. (1)–(3), it is subject to the same restrictions. The most severe limitations implied by (7) are found for optical depths $1 \lesssim \tau \lesssim 10$. There, $t_{diff}$ is already a useful approximation of $t_{rad}$, whence $\Delta t \lesssim h^2/\chi$ as for the heat equation. Consequently, a grid refinement by a factor of 4 in that region requires a 16 times smaller time step. This explains the extremely small $\Delta t$ of 5 ms and 3.4 ms for our 2D simulations with solar helium abundance and without helium, respectively. We note that the coarse grid time steps of about 0.08 s and 0.05 s are comparable to the $\lesssim 0.2$ s reported in Freytag & Steffen (2004) — these differences in $\Delta t$ show that the 2D simulations presented here require computational efforts just slightly smaller than state-of-the-art 3D simulations at lower resolution $h$. But are such small $\Delta t$ unavoidable, if $Q_{rad}$ is computed on the same mesh as the hydrodynamical variables? To check this we have computed the evolution time scale of the independent hydrodynamical variables

$\rho$ and $e$. If we know both variables at the grid at a time step $n$, i.e., $\rho^{(n)}$ and $e^{(n)}$, for any integration method we require that $\Delta t < C\rho^{(n)}(\partial\rho^{(n)}/\partial t)^{-1}$ and $\Delta t < Ce^{(n)}(\partial e^{(n)}/\partial t)^{-1}$, where $C$ is a constant less than 1, typically 0.1, to be able to predict the new state of the system at the time step $n+1$ (otherwise, even with fully implicit methods, iterative solvers may not converge, instabilities can occur, etc.). If we take $C = 0.1$, this is equivalent to requiring that at each grid cell the solution should not change by more than 10% during an integration step. Inspecting the simulation at three subsequent time steps we can easily estimate the time derivative and evaluate these inequalities. We have done this for the first time steps of the simulation and for a number of time steps spread over the entire duration of the run for solar helium abundance. To interpret the results we have also evaluated the convective flow time scale $t_c = h/\max((\mathbf{u}^2)^{1/2})$ as well as $t_s = h/\max c_s$, and finally $\Delta t < Ce^{(n)}(Q_{rad}^{(n)})^{-1}$ to see, if $Q_{rad}$ operates on a shorter time scale than any of the mechanical terms in Eq. (3).

Table 1: Time scales of the simulations with solar helium abundance. Grid 1 denotes the entire coarse grid, grid 2 the fine grid, grid 3 the coarse grid without the domain for which the fine grid is available. $C = 0.1$ and $D = 0.25$ in all calculations.

| time scale | grid 1 | grid 2 | grid 3 |
|---|---|---|---|
| | $t = 0.06$ s | | |
| $Dt_s$ | 1.27 | 0.51 | 1.27 |
| $Dt_c$ | 18673.50 | 2204.89 | 36166.30 |
| $C\min(\rho/\rho_t)$ | 1.48 | 31.11 | 1.48 |
| $C\min(e/e_t)$ | 0.56 | 0.12 | 1.49 |
| $C\min(e/Q_{rad})$ | 0.40 | 0.12 | 17.97 |
| | $t = 3.62$ s | | |
| $Dt_s$ | 1.27 | 0.51 | 1.27 |
| $Dt_c$ | 371.56 | 23.12 | 587.73 |
| $C\min(\rho/\rho_t)$ | 25.60 | 7.60 | 25.60 |
| $C\min(e/e_t)$ | 7.33 | 2.91 | 20.33 |
| $C\min(e/Q_{rad})$ | 0.84 | 2.15 | 64.36 |
| | $t = 11442.92$ s ($\sim 8.3\ t_{sc}$) | | |
| $Dt_s$ | 1.27 | 0.51 | 1.27 |
| $Dt_c$ | 1.86 | 0.49 | 8.05 |
| $C\min(\rho/\rho_t)$ | 0.80 | 0.53 | 6.30 |
| $C\min(e/e_t)$ | 0.81 | 0.53 | 7.35 |
| $C\min(e/Q_{rad})$ | 0.40 | 1.12 | 13.59 |

In Table 1 we show the results for the third time step, for a time step after initial relaxation through radiation, and for a time step during the generation of shock fronts described in the previous subsection. Time steps between the second and the third example show a rather continuous transition between these two. We note that only during the first time steps the radiative heating and cooling $(e/Q_{rad})$ constrains the temporal evolution by providing the shortest time scale (even, if $C = D$; here, $D$ denotes the constant in the Courant-Friedrichs-Lewy condition for the numerical time integration scheme due to sound waves and flow speed as given by $t_s$ and $t_c$ for a given grid size $h$). However, already after a few seconds the temperature perturbations have been smoothed out to an extent that sound speed and its associated time scale $t_s$ set the restrictions for the time evolution of the dynamical variables of the system. In the last time step shown supersonic velocities in the photosphere finally have led to $t_c < t_s$. We conclude that except for the first few hundred simulation time steps totalling just a few seconds, which are subject to the assumed initial conditions and random perturbations applied to the latter, the temporal evolution of the system is governed by the hydrodynamical time scales. This holds even for $Q_{rad}$ itself, which implies changes on the total energy on a similar time scale as hydrodynamical processes (note that the overestimation of $Q_{rad}$ on the coarse grid leads to a slightly shorter time scale, but this part of the solution is discarded by using the fine grid solution — outside the fine grid domain the coarse grid solution imposes no restrictions).

Thus, mathematically $Q_{rad}$ is just a stiff term in a differential equation making the whole problem at least in principle solvable on the evolution timescale of the dynamical variables of the system by a properly designed implicit integration method. The restrictions in $\Delta t$ are solely due to high wave number components $k$ contained in $Q_{rad}$, which represent radiative transfer over one or a few grid cells. This transfer indeed occurs on short time scales $t_{rad}$, but does not govern the evolution of the system itself, as it takes the much longer time $Ce/Q_{rad}$ to substantially change the energy content of a grid cell radiatively. A natural approach is to consider an operator splitting technique and integrate only $Q_{rad}$ by an implicit method. This strategy is also used in simulations of convection in rotating spherical shells to model the lower part of the solar convection zone (Clune et al., 1999). But here the difficulty is that the coefficients in $Q_{rad}$ are neither constant, nor linearly dependent on $(\rho, e)$. Plain subcycling for the integration of $Q_{rad}$ (multiple radiative transfer steps per hydrodynamical time step) alone is inefficient, since $Q_{rad}$ as a whole does not cause rapid changes to $e$, but only its high $k$ components evolve that way, while a consistent update of its coefficients is non-trivial. Thus, we consider it more promising to analyse higher order methods with proper damping of small but rapidly evolving components or filtering methods for their suitability to accelerate RHD simulations of stellar convection at high resolution. This would bring 3D simulations resolving the radiative heating and cooling at the surface of mid A-type main sequence stars from a supercomputing application to the realm of high performance department computers, as have been used in solar granulation simulations with ANTARES (see Muthsam et al. 2007, 2009).

## 6.  Conclusions and outlook

We have presented 2D RHD simulations performed with the ANTARES code for a mid A-type star for two extreme cases of helium abundance, a solar one and a helium free composition. The simulations differ from previous works by a higher resolution and the application of high ($5^{th}$) order advection schemes running stably for these simulations without the need to introduce artificial diffusion (see Muthsam et al. 2009 for further details). The quality of the advection scheme was demonstrated by the need to introduce artificial damping of vertical velocities over several sound-crossing time scales of the simulations to remove vertical oscillations introduced by the initial conditions. After that phase oscillations driven by the flow itself are found to reappear. The mean structure of the convection zone is close to a radiative one, as found in previous RHD simulations in 2D and 3D. The most interesting differences between the two cases we have considered include an inversion of the sign of the flux of kinetic energy as well as higher velocities and larger flow structures, each of them observed for the case with zero helium abundance. This can be understood in terms of the smaller extent of the convection zone due to the absence of partial ionisation of helium and the lower mean molecular weight of a helium free mixture. Since the evolution time scales of the surface convection zone are sufficiently short, we expect these results to be robust. As a note of caution we point out the lack of strong vortices found for the zone of He II ionisation for the case with solar abundance, which is in contrast to the results reported in Freytag (1995) and Freytag et al. (1996). From further simulation runs with different initial conditions we have found that at least for the lower (He II ionisation driven) convection zone the simulations should be performed over very long times to ensure they no longer depend on the initial perturbations (vortices seeded initially may just remain while they may take a long time to grow on their own). As soon as the surface convection zone is sufficiently developed large shock fronts are emitted into the photosphere. These result from the interaction of vortex pairs with the density inversion near the surface. We expect that a similar mechanism could work around strong downflow regions in 3D simulations as well. The large shock fronts might affect spectral line profiles and provide a mechanism of heating chromospheres in mid A-type stars. Finally, we have analysed the importance of resolution to properly compute the radiative heating and cooling rate $Q_{\text{rad}}$ and have shown that the time step restrictions on a hydrodynamical solver implied by computing this quantity on a fine mesh are just those of a classical stiff term in a differential equation. Properly designed implicit integration methods should thus be able to accelerate RHD simulations of this class of stars.

This is most important since the two simulations shown here required 8.3 million time steps for the fine grid solutions resulting in a total of $\sim 13,000$ CPU core hours (on POWER5 1.9 GHz CPUs). A 3D RHD simulation at the same resolution $h$ is hence a supercomputing project (we recall that this is a consequence of the $\Delta t \lesssim h^2/\chi$ dependence of an explicit solver — a larger $h$ as used in previous works would make the simulation more affordable, but this should be traded in only, if $Q_{\text{rad}}$ can be computed sufficiently accurately). Such a project is feasible with the ANTARES code which has successfully been run with good scaling on up to 1024 CPU cores at the POWER6 machine of the RZG in Garching. However, this would still restrict us to very few simulation runs. We thus think it is important to work on the implementation of proper integration methods, since this would also provide benefits for simulations of other types of stars once high enough resolution is demanded. Moreover, it would bring 3D RHD simulations of mid A-type stars with a resolution $h$ as presented here into the realm of high performance department and university computers, which is already the case for equivalent simulations of F-type to M-type main sequence stars.

Concerning the interpretation of our present simulations with respect to the fact that they have been done for two spatial dimensions we consider a comparison with the solar case useful, where high-resolution 2D simulations (Muthsam et al., 2007) were followed by their 3D counterparts somewhat later on (Muthsam et al., 2009). Our present results should be equally valid for the 3D case with respect to our discussion of resolution, of radiative cooling and the stiffness property of $Q_{\text{rad}}$, and the low intrinsic numerical damping of oscillations of the numerical method (the operators in the numerical scheme are one-dimensional and act separately in each spatial direction). We expect pulsational instabilities to be present both in 2D and 3D, as the nature of the driving is not related to the presence of two horizontal directions, although quantitatively the results will change. Since the density inversion is independent of the number of horizontal directions as well, we expect shock fronts to be formed also in the 3D case, but the detailed mechanism might be different (compare Muthsam et al. 2007 with Muthsam et al. 2009). The same holds for other results which directly depend on the flow dynamics: the size and sign of kinetic energy and enthalpy fluxes, the extent of overshooting, the mixing between the two convection zones in the case with solar helium abundance, and the lifetime of structures as well as their detailed dynamical properties. These processes are sensitive to the different dynamics in two and three spatial dimensions and quantitative comparisons with data from spectroscopy and asteroseismology will eventually require 3D simulations of equally high resolution.

We are working on a version of ANTARES with open vertical boundary conditions which avoids the reflection of waves and shock fronts. This will make the simulations more suitable for the study of convection-pulsation interactions and increase the stability of the code, since it avoids extreme conditions occurring when a front hits the upper boundary. We intend to use long-term simulations with this new version of ANTARES to compute synthetic spectra for comparisons with observations, study convection-pulsation interactions, and probe non-local models of convection. This will be followed by 3D simulations of equally high resolution as a final step.

**Acknowledgments.** We thank Ch. Stütz for computing several 1D model atmospheres with the LLmodels code which were used as initial conditions and also thank B. Löw-Baselli for useful discussions on grid refinements. F. Kupka and J. Ballot are grateful to the MPI for Astrophysics and the RZG in Garching for granting access to the IBM POWER5 p575, on which the numerical simulations presented in this paper have been performed. J. Ballot acknowledges support through the ANR project Siroco and H.J. Muthsam is grateful for support through the FWF project P18224-N13. We thank the anonymous referee for remarks which have helped to improve this article.

## References

Adelman, S.J. 2004, in The A-Star Puzzle (IAU Symp. 224), J. Zverko, W.W. Weiss, J. Žižňovský, S.J. Adelman (eds.), (Cambridge, UK: Cambridge University Press), 1

Canuto V. M., & Mazzitelli, I. 1991, ApJ 370, 295

Canuto V. M., & Dubovikov, M. S. 1998, ApJ 493, 834 (CD98)

Canuto V.M., Cheng Y., & Howard A. 2001, J. Atmos. Sci. 58, 1169

Chapman, S. 1954, ApJ 120, 151

Clune, T. C., Elliott, J. R., Miesch, et al. 1999, Parallel Computing 25, 361

Dravins, D. 1987, A&A 172, 211

Dravins, D. 1990, A&A 228, 218

Dravins, D., & Lind, J. 1984, in *Small-Scale Dynamical Processes in Quiet Stellar Atmospheres*, S.L. Keil (ed.) (Sacramento Peak, NM: National Solar Observatory), 414

Ferguson, J.W., Alexander, D.R., Allard, F., et al. 2005, ApJ 623, 585

Freytag, B. 1995, *PhD thesis*, Universität Kiel, Germany

Freytag, B., Steffen, M., & Ludwig, H.-G. 1996, A&A 313, 497

Freytag, B., & Steffen, M. 2004, in *The A-Star Puzzle (IAU Symp. 224)*, J. Zverko, W.W. Weiss, J. Žižňovský, S.J. Adelman (eds.), (Cambridge, UK: Cambridge University Press), 139

Gigas, D. 1984, in *Solar and Stellar Granulation*, R.J. Rutten, G. Severino (eds.) (Dordrecht, Kluwer), NATO Adv. Sci. Inst. (ASI) Serives C, Vol. 263, 533

Gray, D.F. 1988, *Lectures on Spectral-line Analysis: F, G and K Stars*, (Arva, ON, Canada: The Publisher)

Gray, D.F. 1989, PASP, 101, 832

Gray, D.F., & Toner, C.G. 1986, PASP, 98, 499

Gray, D.F., & Nagel, T. 1989, ApJ, 341, 421

Grevesse, N., & Noels, A. 1993, in *Origin and Evolution of the Elements*, N. Prantzos, E. Vangioni-Flam and M. Cassé (eds.), (Cambridge, UK: Cambridge University Press), 15

Gulliver, A.F., Hill, G., & Adelman, S.J. 1994, ApJ 429, L81

Hill, G., Gulliver, A.F., & Adelman, S.J. 2004, in *The A-Star Puzzle (IAU Symp. 224)*, J. Zverko, W.W. Weiss, J. Žižňovský, S.J. Adelman (eds.), (Cambridge, UK: Cambridge University Press), 35

Iglesias, C.A., & Rogers, F.J. 1996, ApJ 464, 943

Kippenhahn, R., & Weigert, A. 1994, *Stellar Structure and Evolution*, 3rd printing (Springer-Verlag, New York)

Kochukhov, O., Freytag, B., Piskunov, N., & Steffen, M. 2007, in *Convection in Astrophysics (IAU Symp. 239)*, F. Kupka, I.W. Roxburgh, K.L. Chan (eds.), (Cambridge, UK: Cambridge University Press), 68

Kupka, F. 2005, in *Element Stratification in Stars: 40 Years of Atomic Diffusion*, G. Alecian, O. Richard, S. Vauclair (eds.), EAS Publications Series, Vol. 17, 177

Kupka, F. 2009, in *Interdisciplinary Aspects of Turbulence*, W. Hillebrandt, F. Kupka (eds.) (Springer Verlag, Berlin), Lecture Notes in Physics Vol. 756, 49

Kupka, F., Paunzen, E., Iliev, I.Kh., & Maitzen, H.M. 2004, MNRAS 352, 863

Kupka, F., & Montgomery, M.H. 2002, MNRAS 330, L6

Kurucz, R.L. 1993, Opacities for Stellar Atmospheres: [+0.0], [+0.5], [+1.0], Kurucz CD-ROM No. 2 (Cambridge, Mass., Smithsonian Astrophysical Observatory)

Kurucz, R.L. 1993, Opacities for Stellar Atmospheres: [-5.0], [+0.0,noHe], [-0.5,noHe], Kurucz CD-ROM No. 8 Cambridge, Mass., Smithsonian Astrophysical Observatory)

Landstreet, J.D. 1998, A&A 338, 1041

Landstreet, J.D. 2004, in The A-Star Puzzle (IAU Symp. 224), J. Zverko, W.W. Weiss, J. Žižňovský, S.J. Adelman (eds.), (Cambridge, UK: Cambridge University Press), 423

Landstreet, J.D. 2007, in Convection in Astrophysics (IAU Symp. 239), F. Kupka, I.W. Roxburgh, K.L. Chan (eds.), (Cambridge, UK: Cambridge University Press), 103

Landstreet, J.D., Kupka, F., Ford, H.A., et al. 2009, A&A, in print (see also arXiv:0906.3824v1 [astro-ph.SR])

Latour, J., Toomre, J., & Zahn, J.-P. 1981, ApJ 248, 1081

Lesieur, M. 1997, Turbulence in Fluids, 3rd edn. (Kluwer Academic Publishers, Dordrecht)

Liu, X., Osher, S., & Chan, T. 1994, J. Comp. Phys. 115, 200

Michaud, G. 2004, in The A-Star Puzzle (IAU Symp. 224), J. Zverko, W.W. Weiss, J. Žižňovský, S.J. Adelman (eds.), (Cambridge, UK: Cambridge University Press), 173

Moeng, C.-H., & Rotunno, R. 1990, J. Atmos. Sci. 47, 1149

Muthsam, H.J., Löw-Baselli, B., Obertscheider, Chr., et al. 2007, MNRAS 380, 1335

Muthsam, H.J., Kupka, F., Löw-Baselli, B., et al. 2009, NewA, submitted (see also arXiv:0905.0177v1 [astro-ph.SR])

Nordlund, Å., & Stein, R.F. 1991, in Stellar Atmospheres: Beyond Classical Models, Proceedings of the Advanced Research Workshop, Trieste, Italy, L. Crivellari et al. (eds.) (Dordrecht, D. Reidel Publishing Co.), 263

Obertscheider, C. 2007, PhD thesis, Universität Wien, Austria

Richard, O., Michaud, G., & Richer, J. 2001, ApJ 558, 377

Rogers, F.J., Swenson, F.J., & Iglesias, C.A. 1996, ApJ 456, 902

Rosenthal, C.S., Christensen-Dalsgaard, J., Nordlund, Å., et al. 1999, A&A 351, 689

Royer, F., Zorec, J., & Gómez, A.E. 2007, A&A 463, 671

Shibahashi, H., Gough, D., Kurtz, D.W., & Kambe, E. 2008, PASJ 60, 63

Shulyak, D., Tsymbal, V., Ryabchikova, T., et al. 2004, A&A 428, 993

Siedentopf, H. 1933, Astron. Nachr. 247, 297

Silaj, J., Townshend, A., Kupka, F., et al. 2005, in *Element Stratification in Stars: 40 Years of Atomic Diffusion*, G. Alecian, O. Richard, S. Vauclair (eds.), EAS Publications Series, Vol. 17, 345

Simon, T., Ayres, T.R., Redfield, S., & Linsky, J.L. 2002, ApJ 579, 800

Sofia, S., & Chan, K.L. 1984, ApJ 282, 550

Spiegel, E.A. 1957, ApJ 126, 202

Steffen, M. 2007, in *Convection in Astrophysics (IAU Symp. 239)*, F. Kupka, I.W. Roxburgh, K.L. Chan (eds.), (Cambridge, UK: Cambridge University Press), 36

Steffen, M., Freytag, B., & Ludwig, H.-G. 2006, in *Proceedings of the 13th Cambridge Workshop on Cool Stars, Stellar Systems and the Sun, held 5-9 July, 2004 in Hamburg, Germany* F. Favata, G.A.J. Hussain, and B. Battrick (eds.) (European Space Agency, SP-560), 985

Trampedach, R. 1997, *Master thesis*, Aarhus University, Denmark

Trampedach, R. 2004, in *The A-Star Puzzle (IAU Symp. 224)*, J. Zverko, W.W. Weiss, J. Žižňovský, S.J. Adelman (eds.), (Cambridge, UK: Cambridge University Press), 155

Vauclair, G., Vauclair, S., & Pamjatnikh, A. 1974, A&A 31, 63

Xiong, D.-R. 1990, A&A 232, 31

Comm. in Asteroseismology
Volume 160, October 2009
© Austrian Academy of Sciences

# Survey of Candidate Pulsating Eclipsing Binaries - I

S. Dvorak[1]

[1] Rolling Hills Observatory, Clermont, FL USA

## Abstract

Initial results from a photometric survey of stars selected from the list of eclipsing binaries that may contain a pulsating component by Soydugan et al. (2006) are reported. A minimum of two nights of CCD observations with V and/or B filters of each of the 35 stars from this list was collected. Of the 35 stars studied, a pulsating component was detected in three of the systems. Pulsations were also serendiptiously detected in the eclipsing binary RR Leporis, which is not on the candidate list.

Accepted:    2009, August 21
Individual Objects:   SZ Ari, CG Aur, V0417 Aur, UW Boo, V364 Cas, V Crt, MY Cyg, V456 Cyg, V466 Cyg, V477 Cyg, SZ Her, TX Her, UX Her, KW Hya, UU Leo, VZ Leo, WY Leo, RR Lep, SX Lyn, BO Mon, EP Mon, V501 Oph, EY Ori, FT Ori, V536 Ori, TY Peg, BG Peg, BO Peg, IQ Per, IU Per, ZZ Pup, AC Tau, AQ Tau, EW Tau, RS Tri, KP Vir.

## 1. Introduction

Soydugan et al. (2006) introduced a catalog of eclipsing binaries with photometric color or spectroscopic classification that placed one or both components within the $\delta$ Scuti region of the instability strip. These stars were selected from existing catalogs of photometric and spectroscopic data, including a number of little-studied stars discovered by the Hipparchos mission.

Eclipsing binaries, particularly those with total eclipses, are especially useful for testing stellar models. Photometry in different bandpasses combined with radial velocity measurements often make it possible to determine accurate physical parameters for the components. These values can then be compared to star models as a check on the model's accuracy. This check is particularly interesting for pulsating variables where the models have to consider the star's internal structure in more detail. In an attempt to find suitable targets for further study a survey of eclipsing binaries possibly containing $\delta$ Scuti-type variables was undertaken at Rolling Hills Observatory (RHO), using the catalog of Soydugan et al. as a source of targets.

## 2.  Procedures

### 2.1.  Selection of Targets

A selection of stars suitable for observations with the 0.25m telescope and CCD at RHO was taken from this catalog. Stars brighter than $V \approx 12$ with declinations between -20 and +57 degrees were selected from this list and examined to ensure that suitable comparison stars were available. From the stars that met these criteria 35 were chosen for the first series of observations reported in this paper.

One additional target, RR Leporis, was added to the list selected from Soydugan et al. This star's short-period oscillations were detected during an earlier, unrelated time series taken while observing a primary eclipse of that star.

### 2.2.  Observation and reductions

$\delta$ Scuti stars typically have periods ranging from approximately 0.2 to 4 hours so each star was observed for a minimum of 4 hours per night, on at least two nights. The amplitude of pulsations are larger in shorter wavelengths so a B filter was used where target and comparison star brightness permitted (roughly $< 8.5$ mag). Fainter stars were measured with a V filter. Time series were generally collected between primary and secondary eclipses, near phases 0.25 and 0.75, to minimize the light variations due to the eclipsing nature of the systems.

All data collection was done with the Meade 0.25m Schmidt-Cassegrain telescope and SBIG ST-9XE CCD at RHO, using either a Johnson B or V filter. Image cadence was typically between 35 and 60 seconds, depending on the target's brightness, as a compromise between photometric accuracy and the need for frequent sampling for the short-period oscillations. Where possible, multiple comparison stars were used to reduce the average error. The precision of individual measurements was generally better than 6 millimags.

The images were processed using normal dark and flat frame processing, and star lists were extracted using aperture photometry with sextractor (Bertin & Arnout 1996). Differential magnitudes for the target and check stars were produced using proprietary software. The nightly differential magnitude measurements for each star was flattened using a polynomial fit to eliminate variations due to the eclipsing nature of the systems. Each night's data was shifted to bring the average of each data set into agreement. Data taken during primary or secondary eclipse where the target star's brightness differed from its out of eclipse brightness by more than approximately 0.3 mag were omitted.

## 3.   Analysis and results

A period search was conducted with the Fourier analysis routine in Period04 (Lenz & Breger 2005) on the normalized data for each star. Of the 35 stars selected from the Soydugan list, three showed statistically significant signals with $S/N > 5$ in the range $6 \geq f \leq 50$ cd$^{-1}$ that is expected for $\delta$ Scuti stars. The results for all 36 stars are shown in Table 1.

Table 1: Observed Stars

| Star | Nights | Freq(cd$^{-1}$) | Comments |
|------|--------|-----------------|----------|
| SZ Ari | 3 | - | - |
| CG Aur | 3 | - | - |
| V0417 Aur | 5 | - | - |
| UW Boo | 4 | - | - |
| V364 Cas | 8 | - | Equal components |
| V Crt | 4 | - | - |
| MY Cyg | 3 | - | - |
| V456 Cyg | 2 | - | - |
| V466 Cyg | 2 | - | - |
| V477 Cyg | 3 | - | - |
| SZ Her | 4 | - | - |
| TX Her | 8 | - | - |
| UX Her | 2 | - | - |
| KW Hya | 2 | - | - |
| UU Leo | 5 | - | - |
| VZ Leo | 5 | - | - |
| WY Leo | 27 | 15.2528(1) | - |
| RR Lep | 13 | 31.8654(3) | - |
| SX Lyn | 4 | - | - |
| BO Mon | 4 | - | - |
| EP Mon | 5 | - | - |
| V501 Oph | 2 | - | - |
| EY Ori | 7 | - | - |

Table continued on next page

continued from previous page

| Star | Nights | Freq(cd$^{-1}$) | Comments |
|------|--------|-----------------|----------|
| FT Ori | 6 | - | - |
| V536 Ori | 4 | - | - |
| TY Peg | 4 | - | - |
| BG Peg | 16 | 24.9857(1) | - |
| BO Peg | 3 | - | - |
| IQ Per | 6 | - | Min II is total |
| IU Per | 2 | - | - |
| ZZ Pup | 5 | - | - |
| AC Tau | 9 | 17.533(1) | - |
| AQ Tau | 3 | - | - |
| EW Tau | 6 | - | - |
| RS Tri | 4 | - | - |
| KP Vir | 15 | - | Unsolved Hipparchos EB |

Table 2 presents information on the four systems containing pulsating components found in this study. The table includes observational information including the target star position and comparison and check stars used, as well as physical parameters such as spectral class and V magnitude obtained from the SIMBAD database.

Table 2: Pulsating Star Details

| Star | Position (J2000) | V mag | Sp Type | Comparison Stars |
|------|------------------|-------|---------|------------------|
| WY Leo | 09:31:01.1 +16:39:25.2 | 11.0 | A2 | GSC 01403-00795 |
| | | | | GSC 01403-01412 |
| | | | | GSC 01403-00075 |
| RR Lep | 05:12:10.5 -13:11:58.6 | 9.98 | A4 | GSC 05342-00022 |
| BG Peg | 22:52:47.2 +15:39:09 | 10.7 | A2 | GSC 01698-00071 |
| AC Tau | 04:37:06.4 +01:41:31.2 | 10.5 | A8 | GSC 00082-00234 |
| | | | | GSC 00083-00652 |
| | | | | GSC 00083-00713 |
| | | | | GSC 00083-00681 |
| | | | | GSC 00083-00483 |

Table 3: Orbital and Pulsation Periods

| Star | Amplitude (V mag) | $P_{orb}$ (days) | $P_{puls}$ (days) |
|------|-------------------|------------------|-------------------|
| WY Leo | 0.011(1) | 4.98578 | 0.0655617(9) |
| RR Lep | 0.005(1) | 0.91543 | 0.0313820(6) |
| BG Peg | 0.015(2) | 1.95243 | 0.0400229(3) |
| AC Tau | 0.006(1) | 2.04340 | 0.057035(7) |

Detailed information on the four systems where pulsations were detected are shown in Table 3. The amplitude from Period04, eclipsing binary orbital period, pulsational period from this study, and the ratio of pulsation to orbital period are shown in the Amplitude, $P_{orb}$, $P_{puls}$, and $P_{puls}/P_{orb}$ columns, respectively. Errors from the least-squares analysis in Period04 are included in parentheses for the amplitude and $P_{puls}$ values. Figures 1 through 4 show the power spectrum for each of the four systems. Representative single-night unflattened light curves for these systems are shown in Figures 5 through 8.

The three positive detections out of 35 targets equals a success rate of 8.6%. All of the systems where pulsations were detected consisted of components that are quite unequal in brightness which made detecting pulsations, presumably in the brighter component, far easier. In systems where the components are close in brightness, such as V364 Cas, or where the fainter star is the pulsator, the detection of pulsations would necessarily be more difficult. Of the 32 systems where no pulsations were detected it is certainly possible that a pulsating component does in fact exist but the pulsations were too small to be detected by the instruments used in this study. In addition, $\delta$ Scuti stars often exhibit multiple, interfering pulsation frequencies, and additional monitoring of these systems might detect pulsations when these frequencies add constructively.

Liakos, A. & Niarchos, P. (2009) contains a survey of 24 stars similar to the current paper. Three of that paper's stars were also studied in the current paper: UW Boo, V456 Cyg, and V466 Cyg. The results in that paper are in agreement with the current results in that no pulsations were detected in any of these three stars. Liakos & Niarchos detected pulsations in three stars, or 12.5%.

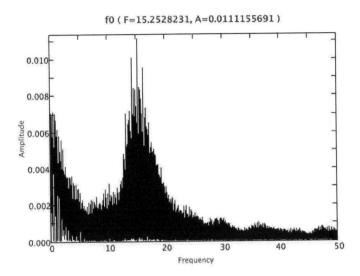

Figure 1: Periodogram of WY Leo

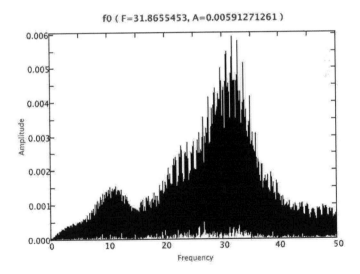

Figure 2: Periodogram of RR Lep

dSct f0 ( F=24.9857068, A=0.0161139259 )

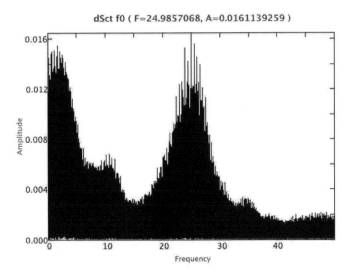

Figure 3: Periodogram of BG Peg

f0 ( F=17.53447, A=0.00527258306 )

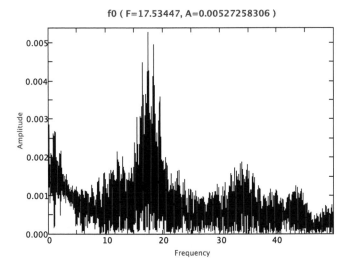

Figure 4: Periodogram of AC Tau

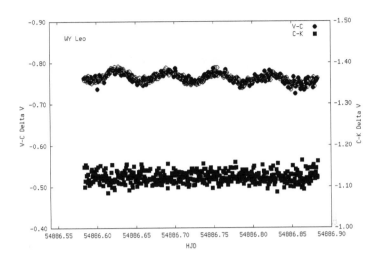

Figure 5: Single-night differential V light curve of WY Leo

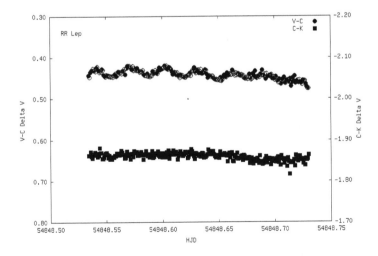

Figure 6: Single-night differential V light curve of RR Lep

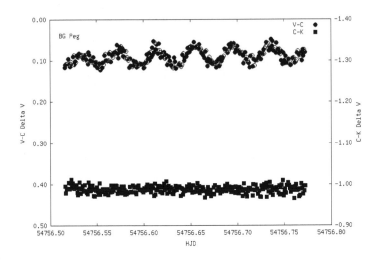

Figure 7: Single-night differential V light curve of BG Peg

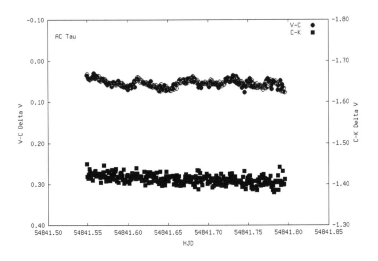

Figure 8: Single-night differential V light curve of AC Tau

## 4.  Conclusions

$\delta$ Scuti stars in eclipsing binary systems can be a useful tool for constraining stellar models.  Of 35 eclipsing binaries in this study that potentially contain such stars selected from *Soydugan et al. (2006)* three (8.6%) were found to have a pulsating component.  The remaining 32 showed no statistically significant variations in the Fourier analysis within the frequency range expected for $\delta$ Scuti stars.  This detection rate is consistent with the 12.5% rate reported by Liakos & Niarchos (2009).  One additional eclipsing system containing a pulsating component, RR Leporis, was previously discovered at RHO and was included in the paper.

**Acknowledgments.**  This research has made use of the SIMBAD database operated at CDS, Strasbourg, France

## References

Bertin, E., & Arnout, S. 1996, A&AS, 117, 393

Lenz, P. & Breger, M. 2005, CoAst, 146, 53

Liakos, A. & Niarchos, P. 2009, CoAst, 160, 2

Soydugan, E., Soydugan, F., Demircan, O., & Ibanoglu, C. 2006, MNRAS, 370, 2013

Comm. in Asteroseismology
Volume 160, October 2009
© Austrian Academy of Sciences

# Automated extraction of oscillation parameters for Kepler observations of solar-type stars

Daniel Huber[1], Dennis Stello[1], Timothy R. Bedding[1], William J. Chaplin[2],
Torben Arentoft[3], Pierre-Olivier Quirion[3] and Hans Kjeldsen[3]

[1] Sydney Institute for Astronomy (SIfA), School of Physics, University of Sydney,
NSW 2006, Australia
[2] School of Physics and Astronomy, University of Birmingham, Edgbaston,
Birmingham, B15 2TT, UK
[3] Department of Physics and Astronomy, University of Aarhus, DK-8000 Aarhus C,
Denmark

## Abstract

The recent launch of the Kepler space telescope brings the opportunity to study oscillations systematically in large numbers of solar-like stars. In the framework of the asteroFLAG project, we have developed an automated pipeline to estimate global oscillation parameters, such as the frequency of maximum power ($\nu_{max}$) and the large frequency spacing ($\Delta\nu$), for a large number of time series. We present an effective method based on the autocorrelation function to find excess power and use a scaling relation to estimate granulation timescales as initial conditions for background modelling. We derive reliable uncertainties for $\nu_{max}$ and $\Delta\nu$ through extensive simulations. We have tested the pipeline on about 2000 simulated Kepler stars with magnitudes of $V \sim 7$–$12$ and were able to correctly determine $\nu_{max}$ and $\Delta\nu$ for about half of the sample. For about 20%, the returned large frequency spacing is accurate enough to determine stellar radii to a 1% precision. We conclude that the methods presented here are a promising approach to process the large amount of data expected from Kepler.

Accepted:    2009, October 10

## 1.　Introduction

Stellar oscillations are a powerful tool to study the interiors of stars and to determine their fundamental parameters. Until recently, the detection of oscillations in solar-type stars has been possible only for a handful of bright stars (see, e.g., Bedding & Kjeldsen 2008). With the launch of the space telescopes CoRoT (Baglin et al. 2006) and Kepler (Borucki et al. 2008), however, this situation is changing. In pursuing its main mission goal of detecting transits of extrasolar planets around solar-like stars, Kepler will photometrically monitor thousands of stars for a period of up to four years. Asteroseismology will allow us to determine radii of exoplanet host stars (Christensen-Dalsgaard et al. 2007; Stello et al. 2007; Kjeldsen et al. 2009), and also to study oscillations systematically in a large number of solar-type stars for the first time.

　　To deal with the amount of data that Kepler is expected to return, automatic analysis pipelines are needed. Such algorithms have already been successfully applied to CoRoT exofield data to study oscillations in red giants (Hekker et al. 2009). For Kepler, the development of analysis tools has been carried out in the framework of the asteroFLAG project (Chaplin et al. 2008a; Mathur et al. 2009) through so-called Hare & Hounds exercises, in which one group (the Hounds) analyse simulated data produced by others (the Hares) without knowing the parameters on which the simulations are based. Chaplin et al. (2008b) presented the results of the first exercise, which concentrated on a few stars simulated at different evolutionary stages with various apparent magnitudes and a time base of 4 years (as expected for a full-length Kepler time series). The results were then used in a second exercise to test the ability to determine radii using stellar models (Stello et al. 2009).

　　In this paper, we describe an automated pipeline to extract oscillation parameters such as the frequency of maximum power ($\nu_{max}$) and the mean large frequency spacing ($\Delta\nu$). We apply it to a large sample of simulated time series that are based on stellar parameters of real stars selected for the Kepler asteroseismology survey phase. During this phase, which will occupy the first nine months of Kepler science operations, about 2000 stars will be monitored for one month each. These data are intended to characterise a large number of solar-like stars and the results will be used to verify the Kepler Input Catalog (Brown et al. 2005), as well as to select high-priority targets to be observed for the entire length of the mission.

## 2.　asteroFLAG simulations

The simulated Kepler light curves were produced using a combination of the asteroFLAG simulator (Chaplin et al., in preparation) and the KASOC simulator

(T. Arentoft, unpublished). All simulations include stellar granulation, activity cycles and instrumental noise, as well as oscillation frequencies computed using the stellar evolution and pulsation codes ASTEC (Christensen-Dalsgaard 2008b) and ADIPLS (Christensen-Dalsgaard 2008a), together with rotational splitting and theoretical damping rates. To simulate Kepler survey targets, fundamental parameters were taken from the Kepler Input Catalog. Next, a model within estimated uncertainties of these parameters was chosen for each star. Every simulated light curve had a length of one month, with a sampling time of 60 seconds (representative for real Kepler time series). In total, 1936 stars in the magnitude range $V \sim 7$–$12$ were simulated, and we used this sample to test the pipeline that is described in the next section.

## 3.    Data analysis pipeline

The pipeline covers the first basic analysis steps that will be performed on the Kepler light curves. These are: (a) estimating the position of power excess in the power spectrum, (b) fitting to and correcting for the background, and (c) estimating the mean large frequency spacing. Locating the power excess due to oscillations not only constrains fundamental parameters of a star (in particular, its luminosity), but is also crucial for a successful automation of subsequent analysis steps. A problem when analysing the oscillation signal is the non-white background noise due to variability caused by granulation and stellar activity. For the analysis of red giants observed with CoRoT, Kallinger et al. (2008b) used simultaneous fitting of the background and the oscillation power excess, with the latter modelled with a Gaussian function. Here, we separate these two steps by first locating the power excess region, and then excluding the identified region when modelling the background. Finally, the background-corrected spectrum is used to estimate the large frequency spacing in the region where the power was located. In the following subsections, each of these three analysis steps will be described in detail.

### 3.1.    Locating the power excess

To locate the power excess, we follow a three-step procedure that is demonstrated in Figure 1 using a 30-day VIRGO time series of the Sun (Frohlich et al. 1997):

(1) The background is crudely estimated by binning the power spectrum in equal logarithmic bins and smoothing the result with a median filter. The optimal width of the bins depends on the frequency resolution of the data, and typical values for the 30-day asteroFLAG stars were logarithmic bins with a width of $0.005 \log(\mu\text{Hz})$.

(2) The residual power spectrum, after subtracting this background (Figure 1, top panel), is divided into subsets roughly equal to $4\Delta\nu$ and overlapping by 50 $\mu$Hz. The mean of each subset is subtracted and the absolute autocorrelation function (ACF) for each is calculated for a pre-defined range of frequency spacings (Figure 1, middle panel). Note that in order to conserve information about the actual power level in the power spectrum of a subset, the ACF is not normalised to unity at zero spacing.

(3) For each subset, represented by its central frequency, we collapse the ACF over all frequency spacings (Figure 1, bottom panel). We finally fit a Gaussian function to the peak of the collapsed ACF to localise the power excess region (thick grey line). We take the centre of the Gaussian to be our measurement of the frequency of maximum power, $\nu_{max}$.

More precisely, a vertical cut through the middle panel of Figure 1 at a given frequency is the ACF of the power spectrum subset centered at that frequency. In this example, the subset length chosen was $4\Delta\nu$ ($\sim 540\,\mu$Hz). In applications where no estimate for $\Delta\nu$ is available, a range of up to three subset widths are applied and the one returning the highest S/N in the collapsed ACF is taken for the $\nu_{max}$ estimate. As expected for this example, the ACF shows large values at multiples of half the large frequency spacing of the Sun ($\sim 68\,\mu$Hz), concentrated at frequencies around 3 mHz in the power spectrum. The collapsed ACF in the bottom panel is calculated by summing the middle panel vertically. For comparison, the dashed line shows the power spectrum smoothed with a Gaussian function with a FWHM of $4\Delta\nu$.

An advantage of this technique over smoothing the power spectrum is that the collapsed ACF is strongly sensitive to the regularity of the peaks, rather than just their strengths. In other words, by applying an autocorrelation we use the information that peaks are expected to be regularly spaced, whereas this information is disregarded when smoothing the power. The single strong peak close to 6 mHz in the top panel of Figure 1, for example, is an artefact in the VIRGO photometry and produces a much more significant response in the smoothed spectrum than in the collapsed ACF. Figure 2 illustrates this further, using a power spectrum of a simulated Kepler star with low signal-to-noise. Compared to the smoothed power spectrum, the collapsed ACF shows a strong peak at the correct location of $\nu_{max}$ and is clearly less sensitive to areas with white noise.

The shape of the power envelope in other stars can be very different than the Sun (e.g. for Procyon, see Arentoft et al. 2008) and hence might not be suitably modelled with a simple single Gaussian function. The collapsed ACF is less influenced by power asymmetries than a smoothed spectrum (see Figure 1, bottom panel), and an extension of the pipeline to include multiple Gaussians or

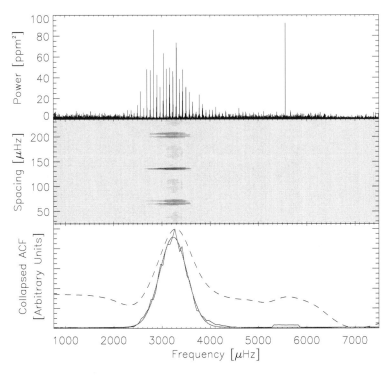

Figure 1: Procedure for locating the power excess using a 30 day subset of VIRGO photometry. Top panel: Background corrected power spectrum. Middle Panel: Auto-correlation as a function of frequency spacing and central frequency of the subset at which the correlation is evaluated. Dark colors are regions of high correlation. Bottom Panel: Collapsed ACF (black solid line) and smoothed power spectrum (dashed line). The grey solid line shows a Gaussian fit to the collapsed ACF.

different functions will be forthcoming. This will be of particular interest when analysing binaries in which both components show detectable power excess in the spectrum.

## 3.2.  Background modelling

Modelling of power due to stellar background is widely done using a sum of power laws initially proposed by (Harvey 1985), with a revision of the power law exponent by Aigrain et al. (2003). Here, we use a mixture of the two versions that was originally suggested by Karoff (2008) and has the form

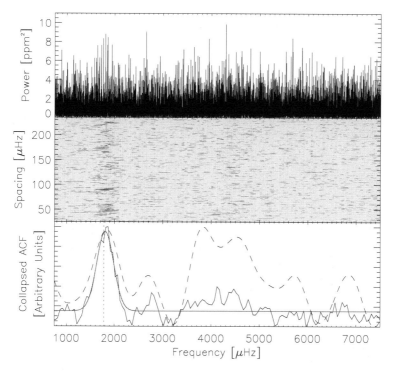

Figure 2: Same as Figure 1 but for a 1-month time series of a simulated Kepler star with low S/N. The vertical dotted line in the bottom panel is the true $\nu_{max}$ value for this simulation. The large spacing was 88 $\mu$Hz.

$$P(\nu) = P_n + \sum_{i=0}^{k} \frac{4\sigma_i^2 \tau_i}{1 + (2\pi\nu\tau_i)^2 + (2\pi\nu\tau_i)^4} , \qquad (1)$$

where $P_n$ is the white noise component, $k$ is the number of power laws used and $\sigma$ and $\tau$ are the rms intensity and timescale of granulation, respectively. The motivation behind this extended model is a more physically realistic interpretation of the stellar background. Instead of assuming a constant slope for the entire frequency range, it allows a shallower slope at low frequencies corresponding to turbulence (stellar activity) and steeper slopes at higher frequencies corresponding to granulation (Nordlund et al.1997).

We determine $\sigma$ and $\tau$ using a least-squares fit to the power density spectrum. From a statistical point of view such an approach is questionable, since a raw power spectrum is not described by Gaussian statistics. We overcome this problem by smoothing the power spectrum using independent averages only (Garcia et al. 2009), which allows a determination of parameter uncertainties. Alternative methods such as a Bayesian approach using Markov-Chain Monte-Carlo simulations (T. Kallinger, private communication) could be used for more detailed studies of stellar granulation.

Regardless of the method of fitting, a pipeline relies on good initial conditions for a successful fit. To estimate such values, it is important to understand the rms intensity and, especially, the timescale of stellar granulation as a function of physical parameters. Based on numerical simulations of stellar surface convection, Freytag & Steffen (1997) initially suggested that the linear size of a granule $l$ is proportional to the pressure scale height on the stellar surface:

$$H_p^{\text{surf}} = \frac{l}{\alpha} .$$    (2)

Here, $\alpha$ denotes the mixing length parameter. Assuming that the cells move proportional to the speed of sound $c_s$ (Svensson & Ludwig 2005), Kjeldsen & Bedding (in preparation) show that, under the further assumption of adiabacity and an ideal gas, the granulation timescale can be expressed as

$$\tau_{\text{gran}} \propto \frac{H_p^{\text{surf}}}{c_s} \propto \frac{L}{T_{\text{eff}}^{3.5} M} .$$    (3)

According to Kjeldsen & Bedding (1995), this is inversely proportional to $\nu_{\text{max}}$ and hence

$$\tau_{\text{gran}} = \tau_{\text{gran},\odot} \frac{\nu_{\text{max},\odot}}{\nu_{\text{max}}} .$$    (4)

This suggests that the timescale of granulation scales with the timescale of oscillations, which is plausible since both processes are tied to convection. Knowing $\nu_{\text{max}}$ from the estimation performed in the previous section, granulation timescales can be scaled from the Sun without prior knowledge of stellar parameters.

We tested this scaling method on HD 49933, a star for which granulation and solar-like oscillations have been detected by CoRoT (Appourchaux et al. 2008). Figure 3 compares the power density spectrum of HD 49933 with the Sun, together with the individual power law components calculated using Equation 1. We assumed three components of stellar background in each spectrum: stellar activity at very low frequencies and two components due to different types of granulation. While the initial guesses for HD 49933 (dashed lines) using Equation 4 yield a satisfactory final fit (dotted and solid lines), it is evident that the background in this star is somewhat different from the scaled Sun. This result confirms that granulation signatures in hotter stars are quite different

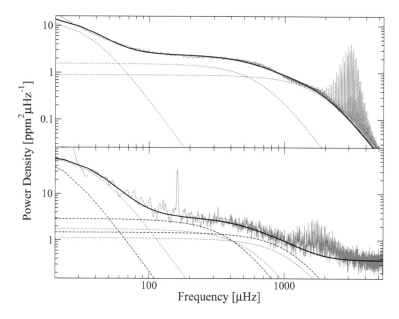

Figure 3: Power density spectra of a 9-year VIRGO time series (top panel) and a 60-day time series of HD49933 as observed by CoRoT (bottom panel). Dotted lines are the fitted individual power law components which, together with a white noise component (not shown), result in the final background model (thick solid lines). Dashed lines in the bottom panel show the initial guesses for the background fit using Equation 4. Note the differences in y-axis scaling for each panel.

from the Sun, as found by Guenther et al. (2008) for a convection model of Procyon. We refer to Ludwig et al. (2009) for a detailed discussion of the granulation signal in HD 49933 in the context of hydrodynamical simulations.

Despite these differences for HD 49933, Equation (4) appears to be a satisfactory approximation to provide initial values for background fitting and hence we implemented it in the pipeline. After the background has been successfully fitted, the power spectrum is corrected by dividing through the background model.

## 3.3.   Estimation of $\Delta \nu$

It is well known that the stochastic excitation and damping of solar-like oscillations causes series of peaks centred around the true frequency values in the power spectrum. To obtain a robust estimate of the average large frequency spacing, it is often helpful to divide the time series into subsets and co-add the corresponding power spectra, which leads to an average power spectrum with lower frequency resolution. In our pipeline, the background-corrected power spectrum is inverse-Fourier-transformed into the time-domain. The time series is then divided up into overlapping subsets (typically of 5 day length with a step size of 1 day) and power spectra of the individual subsets are co-added. This procedure forms a smoothed power spectrum. Figure 4 compares the original power spectrum with the background-corrected co-added power spectrum for a simulated Kepler star.

As a next step, we repeat the power excess determination described in Section 3.1 using the background corrected co-added power spectrum. Using this final value for $\nu_{max}$, we estimate the expected spacing by using the tight correlation between $\nu_{max}$ and $\Delta\nu$ discussed by Stello et al. (2009, in preparation)

$$\Delta\nu_{exp} \propto \nu_{max}^{0.8} .$$  (5)

Next, the autocorrelation of the power spectrum for the region $\nu_{max} \pm 10\Delta\nu_{exp}$ is calculated. Note that this width broadly agrees with the observed power excess in the Sun, and that we are at this stage only interested in deriving an average large frequency spacing over a large number of modes. Finally, we flag the five highest peaks in the autocorrelation, and fit a Gaussian function to the peak among the five which is closest to $\Delta\nu_{exp}$, yielding the final determination of $\Delta\nu$. Figure 5 shows a $\Delta\nu$ measurement of a simulated Kepler star for which the correct spacing does not correspond to the highest peak in the autocorrelation.

## 3.4.    Uncertainties in $\nu_{max}$ and $\Delta\nu$

A crucial part of an automated pipeline is the ability to interpret the quality (or credibility) of the returned values. However, since we determine $\nu_{max}$ and $\Delta\nu$ using least-squares fits to autocorrelation functions, determining reliable uncertainties is not straight forward. As pointed out by Chaplin et al. (2008b), the formal uncertainties of such fits are strongly underestimated since the datapoints to which functions are fitted are highly correlated, and no proper weights (or uncertainties) can be assigned to individual datapoints. Additionally, the stochastic nature of solar-like oscillations introduces an intrinsic scatter of our measured $\nu_{max}$ and $\Delta\nu$ values.

To overcome this, we followed the approach of Chaplin et al. (2007) and performed simulations by producing synthetic time series. The inputs for each simulation were solar frequencies covering roughly twelve orders of $\ell = 0 - 2$ modes taken from BiSON observations (Broomhall et al. 2009). We modelled the amplitudes using a solar envelope derived from smoothing a power spectrum

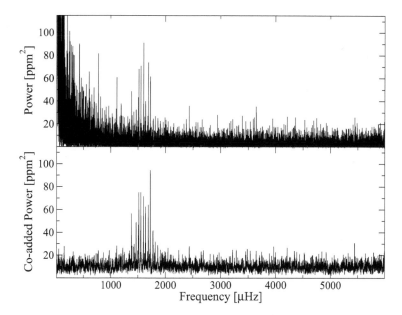

Figure 4: Top panel: Original power spectrum of a simulated Kepler star with a time base of one month. Bottom panel: Background-corrected and co-added power spectrum of the same star.

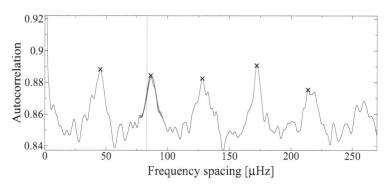

Figure 5: ACF of the background-corrected co-added power spectrum of a simulated Kepler star. Black crosses mark the five highest peaks. The grey line is a fit to the peak among the five which is closest to $\Delta\nu_{\mathrm{exp}}$ (vertical dotted line).

calculated from a 30-day subset of VIRGO photometry. For simplicity, we assumed that all modes are intrinsically equally strong, but accounted for different spatial responses of $\ell = 0 - 2$ modes according to Kjeldsen et al. (2008a). We

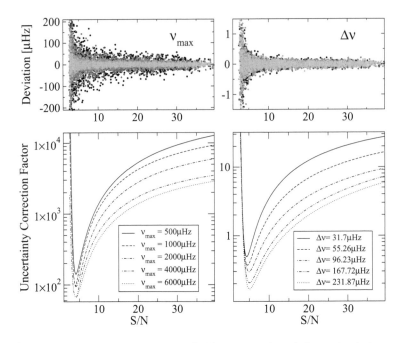

Figure 6: Top panels: Scatter of $\nu_{max}$ (left) and $\Delta\nu$ (right) from simulations as a function of S/N. Darker colours correspond to higher input values of $\nu_{max}$ and $\Delta\nu$, respectively. Bottom panels: Correction factors for formal uncertainties as a function of S/N and input value.

simulated the stochastic excitation and damping using the method of Chaplin et al. (1997), with a frequency independent mode lifetime of three days (i.e. solar). Each time series consisted of the same sampling and time base as the simulated asteroFLAG stars, and white noise was added to each synthetic time series.

We performed simulations with different input amplitudes and frequencies to resemble a range of stellar evolutionary states, and with different S/N corresponding to a variety of stellar magnitudes. We made 100 realizations, including stochastic excitation and white noise for each set of input parameters. The resulting light curves were then analysed by the pipeline, and the standard deviations of the determined values for $\nu_{max}$ and $\Delta\nu$ were taken as the true uncertainties.

The results of the simulations are shown in Figure 6. The top panels show the scatter in $\nu_{max}$ and $\Delta\nu$ as a function of S/N. We see that the scatter of $\nu_{max}$ is greater for higher values of $\nu_{max}$ (darker symbols), while the scatter in $\Delta\nu$ is almost independent of $\Delta\nu$. This is expected, since the power excess hump used to determine $\nu_{max}$ becomes broader for higher input values of $\nu_{max}$ and hence the absolute deviation increases. On the other hand, the peak in the autocorrelation used to determine $\Delta\nu$ will remain about the same because it is determined by the frequency resolution and mode lifetime, which are the same for all simulations. We note that the scatter reaches a constant level for high S/N, and the maximum precision with which $\nu_{max}$ and $\Delta\nu$ can be determined are $\sim 10\,\mu$Hz and $\sim 0.1\,\mu$Hz, respectively. The latter value is in good agreement with the uncertainties reported in the first asteroFLAG exercise (Chaplin et al. 2008b).

The ratio between these values and the formal uncertainties as determined by the least-squares fit give a look up table of correction factors which we use to convert the formal uncertainties into more realistic values. To obtain smoothly varying correction factors, we fitted power laws to the results of the simulations. These are shown in the bottom panels of Figure 6. As expected, the factors increase for higher S/N, i.e. the least squares fit underestimates uncertainties more for higher signal because the correlation between fitted points is higher. Towards the detection limit at low S/N, this trend quickly reverses and formal uncertainties must be scaled with high factors to accommodate the large uncertainty due to high noise levels. We note that at S/N values around 10 and lower, the $\Delta\nu$ uncertainties determined by the least-squares fit are in fact overestimated, with correction factors $< 1$. Considering that our simulations are simplified compared to real data and therefore the scatter at low S/N values is likely underestimated, we disregard this effect and do not downscale formal uncertainties.

We note that the uncertainty correction presented here will also be applicable to real Kepler stars, with slight adaptations for different sampling and observing lengths. In preparation for this, we test our uncertainties using simulated Kepler stars, which will be presented in the next section.

## 4. Application to simulated Kepler observations

We applied the pipeline, as described in the previous section, to 1936 simulated Kepler stars discussed in Section 2. To verify the values returned by the pipeline, we calculated "true" values of $\nu_{max}$ and $\Delta\nu$ as follows: Using the stellar mass, luminosity and effective temperature of the input model, we calculated $\nu_{max,true}$ using the scaling relation by Kjeldsen & Bedding (1995). To determine $\Delta\nu_{true}$, we first determined the input model frequency closest to $\nu_{max,true}$. We then fitted a linear regression to ten orders of the same degree around the frequency of maximum power, and used the slope to estimate the frequency spacing (Kjeldsen et al. 2008b). This was done separately for modes of $\ell = 0 - 2$, and the final value of $\Delta\nu_{\text{true}}$ is a weighted mean of the three spacings, with weights corresponding to the spatial responses as given by Kjeldsen et al. (2008a). Note that for more evolved stars ($\Delta\nu < 70\,\mu$Hz), no reliable model frequencies were available and hence the scaling relation by Kjeldsen & Bedding (1995) was used to calculate $\Delta\nu_{\text{true}}$.

To investigate systematic effects in our uncertainty simulations from section 3.4, we repeated these simulations for noise-free realizations and compared $\nu_{\text{max}}$ and $\Delta\nu$ returned by the pipeline to $\nu_{\text{max,true}}$ and $\Delta\nu_{\text{true}}$ for these simulations. We found that on average the determined $\nu_{\text{max}}$ values are $\sim 1\%$ higher and that the determined $\Delta\nu$ values are $\sim 0.05\,\%$ lower than the input values. Both effects are easily understood: In our uncertainty simulations, as well as the Kepler simulations, a solar oscillation profile was assumed. While the amplitudes in this profile have positive asymmetry, the large spacings increase towards higher frequencies. The collapsed ACF used to determine $\nu_{\text{max}}$ is sensitive to regular peak spacings, and hence overestimates $\nu_{\text{max}}$ compared to our definition of $\nu_{\text{max,true}}$, which is not equal to the center of the solar envelope. The single ACF used to measure $\Delta\nu$ is influenced by the peak power, and hence underestimates $\Delta\nu$ compared to our definition of $\Delta\nu_{\text{true}}$, which is the mean spacing across the envelope independent of amplitude. For this application, we account for both effects by multiplying the measured $\nu_{\text{max}}$ values by 0.99 and the measured $\Delta\nu$ values by 1.0005.

We now proceed to the main results by comparing the measured values of the 1936 Kepler stars with the true values. Figures 7 and 8 display the differences between the true values (as defined above) to the quantities measured by the pipeline for $\nu_{\text{max}}$ and $\Delta\nu$, respectively. In this comparison we have eliminated all stars with relative uncertainties greater than 10%. Furthermore, we disregard extreme outliers by considering only measurements for which the absolute difference of $\Delta\nu$ and $\Delta\nu_{\text{exp}}$ is lower than $0.1\,\Delta\nu_{\text{exp}}$, and the absolute difference of $\nu_{\text{max}}$ and $\nu_{\text{max,exp}}$ is lower than $0.5\,\nu_{\text{max,exp}}$. $\nu_{\text{max,exp}}$ is calculated using the scaling relation by Kjeldsen & Bedding (1995) with the Kepler Input Catalog parameters for these stars. Of the entire sample, 52% of the $\nu_{\text{max}}$ and 48% of the $\Delta\nu$ measurements fulfill these criteria.

The largely symmetric distributions for both $\nu_{\text{max}}$ and $\Delta\nu$ in the top panels of Figures 7 and 8 suggest that systematic effects caused by the methods of the pipeline have been mostly removed or corrected. We suspect that the slight negative trend for high values of $\Delta\nu$ is due to the fact that $\Delta\nu_{\text{true}}$ is calculated based on a fixed number of model frequencies, which underestimates $\Delta\nu_{\text{true}}$ compared to simulated Kepler stars where less low-frequency modes might be visible. As discussed in section 3.4, the scatter in $\nu_{\text{max}}$ gets considerably larger as the power excess shifts to higher frequencies and the oscillation amplitudes become smaller. The bottom panels show the mean uncertainties and the rms

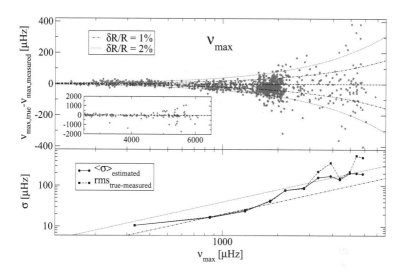

Figure 7: Top panel: Differences between true and measured values for $\nu_{max}$ using all measurements with an uncertainty precision lower than 10% and disregarding extreme outliers (see text). The inset shows the comparisons on a larger scale to display measurements with larger deviations. Bottom panel: Mean estimated uncertainties of the measurements (solid line) compared to the actual scatter of measurements and true values (dashed line). In both panels, the dashed-dotted and dotted lines display uncertainty limits necessary to constrain stellar radii to 1% and 2%, respectively.

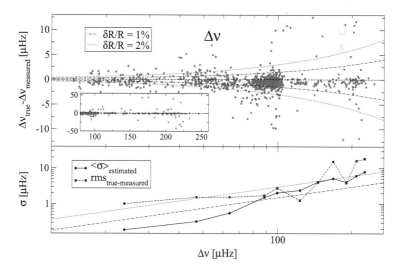

Figure 8: Same as Figure 7 but for $\Delta\nu$.

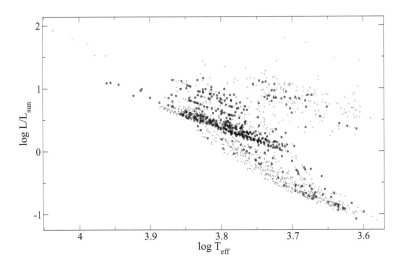

Figure 9: H-R diagram of all 1936 stars in the test sample based on the model parameters. Black circles show stars for which the pipeline returned correct large frequency spacings precise enough to determine the stellar radius to a precision of 1%.

of measured minus true values. In both cases, the curves are for the most part overlapping. The exception are low values of $\Delta\nu$ for which the uncertainties still seem considerably underestimated. We suspect that this is partially connected to the fact that for these stars $\Delta\nu_{true}$ was calculated using a scaling relation rather than the actual model frequencies (see above).

An important application of asteroseismology within the Kepler mission will be to determine radii of of exoplanet host stars. As noted by Chaplin et al. (2008b), a radius determination to a precision of 1% requires a relative uncertainty of 0.15 % on the large frequency spacing. Formally, this corresponds to a 2% relative uncertainty on $\nu_{max}$. These precisions are indicated in Figures 7 and 8 by dashed-dotted lines. We also show the precisions required for a more pessimistic radius precision of 2% (dotted lines). The scatter of the measured values and the mean uncertainties shows that the 1% limit should be achievable for a considerable number of stars up to $\Delta\nu \sim 100\,\mu$Hz and $\nu_{max} \sim 2000\,\mu$Hz, and even higher values if the criterion is relaxed to a radius precision of 2% or 3%. These results therefore indicate an optimistic outlook for the automated analysis of Kepler asteroseismology stars, despite the fact that the time base of the initial survey will only be about 30 days.

To analyse these results in terms of stellar evolution, Figure 9 shows a H-R diagram based on the model parameters of all 1936 sample stars. The majority of the test sample was made up of main sequence and sub-giant stars. As expected, a large part of the best $\Delta\nu$ determinations (with uncertainties allowing a radius determination to 1% precision) were made in stars with relatively high luminosities. The relative lack of detections for the evolved sub-giants compared to the rest of the sample points to a problem of our pipeline with handling signal at the very low frequencies. Quite surprisingly, the results also indicate that precise $\Delta\nu$ determinations will be possible for a number of cool, low-mass main-sequence stars (which represent the datapoints at very high $\nu_{max}$ and $\Delta\nu$ in Figures 7 and 8).

## 5.  Summary & Conclusions

We have described an automated analysis pipeline to extract global oscillation parameters for a large number of stars. We demonstrated that the use of a collapsed autocorrelation function is a sensitive tool to find the location of excess power. We further showed that a determination of $\nu_{max}$ can be used to scale granulation timescales in order to model the background contribution in the power spectrum. To obtain robust uncertainty estimates on $\nu_{max}$ and $\Delta\nu$, we have performed realistic simulations of solar-like oscillations as a function of S/N and $\nu_{max}$, and derive correction factors which are necessary to convert least-squares uncertainties derived from correlated data to realistic uncertainties. Our simulations indicate that for a one-month time series with one-minute sampling, the maximum precision with which $\nu_{max}$ and $\Delta\nu$ can be determined are $\sim$10 $\mu$Hz and $\sim$0.1 $\mu$Hz, respectively.

The automated pipeline was applied to a sample of 1936 simulated stars representing targets of the Kepler asteroseismic survey phase. We show that our scaled uncertainties are reliable for all values of $\nu_{max}$, but seem to be significantly underestimated for $\Delta\nu < 50\mu$Hz. While we suspect that this is mostly due to the fact that $\Delta\nu_{true}$ could not be calculated accurately from model frequencies for evolved stars, the results show that in general some modifications of the code are needed for processing stars that pulsate at low frequencies ($< 500\,\mu$Hz). The further development of the pipeline, in particular with respect to the background modelling, will focus on this adaptation to process red giant stars.

The comparison of real and measured values showed that in 70% and 60% of all cases, $\nu_{max}$ and $\Delta\nu$ were recovered within 10 % of the true value, respectively. Using the estimated uncertainties to eliminate measurements with too large uncertainties and disregarding extreme outliers, these numbers drop to roughly 50%. The scatter of the measured values around the input values and the mean uncertainties agrees well for this sample, and indicate that for at least 20% of the stars $\Delta\nu$ can be determined with a precision sufficiently high to infer stellar radii to 1% accuracy. Plotting these stars in an HR diagram suggests that their distribution is quite diverse, including low-mass main sequence stars as well as evolved sub-giants.

**Acknowledgments.** This work benefited from the support of the International Space Science Institute (ISSI), through a workshop programme award. It was also partly supported by the European Helio- and Asteroseismology Network (HELAS), a major international collaboration funded by the European Commission's Sixth Framework Programme. WJC acknowledges the support of the UK Science and Technology Facilities Council (STFC). DH likes to thank Rafael Garcia, Christoffer Karoff, Stephen Fletcher, Michael Gruberbauer and Thomas Kallinger for interesting and fruitful discussions.

## References

Aigrain, S., Gilmore, G., Favata, F. & Carpano, S. 2003, in Astronomical Society of the Pacific Conference Series, 294, Scientific Frontiers in Research on Extrasolar Planets, ed. D. Deming & S. Seager, 441

Appourchaux, T., Michel, E., Auvergne, M., et al. 2008, A&A, 488, 705

Arentoft, T., Kjeldsen, H., Bedding, T. R., et al. 2008, ApJ, 687, 1180

Baglin, A., Michel, E., Auvergne, M., & The COROT Team 2006, in ESA Special Publication, 624, Proceedings of SOHO 18/GONG 2006/HELAS I, Beyond the spherical Sun

Bedding, T. R., & Kjeldsen, H. 2008, in Astronomical Society of the Pacific Conference Series, 384, 14th Cambridge Workshop on Cool Stars, Stellar Sysems, and the Sun, ed. G. van Belle, 21

Borucki, W., Koch, D., Basri, G., et al. 2008, in IAU Symposium, 249, IAU Symposium, ed. Y.-S. Sun, S. Ferraz-Mello, & J.-L. Zhou, 17

Broomhall, A.-M., Chaplin, W. J., Davies, G. R., et al. 2009, MNRAS, 396, L100

Brown, T. M., Everett, M., Latham, D. W., & Monet, D. G. 2005, in Bulletin of the American Astronomical Society, 37, Bulletin of the American Astronomical Society, 1340

Chaplin, W. J., Appourchaux, T., Arentoft, T., et al. 2008a, JPhCS, 118, 012048

Chaplin, W. J., Appourchaux, T., Arentoft, T., et al. 2008b, AN, 329, 549

Chaplin, W. J., Elsworth, Y., Howe R., et al. 1997, MNRAS, 287, 51

Chaplin, W. J., Elsworth, Y., Miller, B. A., et al. 2007, ApJ, 659, 1749

Christensen-Dalsgaard, J. 2008a, Ap&SS, 316, 113

Christensen-Dalsgaard, J. 2008b, Ap&SS, 316, 13

Christensen-Dalsgaard, J., Arentoft, T., Brown, T. M., et al. 2007, CoAst, 150, 350

Freytag, B., & Steffen, M. 1997, in Astronomische Gesellschaft Abstract Series, 13, Astronomische Gesellschaft Abstract Series, ed. R. E. Schielicke, 176

Frohlich, C., Andersen, B. N., Appourchaux, T., et al. 1997, SoPh, 170, 1

Garcia, R. A., Regulo, C., Samadi, R., et al. 2009, ArXiv e-prints

Guenther, D. B., Kallinger, T., Gruberbauer, M., et al. 2008, ApJ, 687, 1448

Harvey, J. 1985, High-resolution helioseismology, Tech. rep.

Hekker, S., Kallinger, T., Baudin, F., et al. 2009, ArXiv e-prints

Kallinger, T., Gruberbauer, M., Guenther, D. B., et al., 2008a, ArXiv e-prints

Kallinger, T., Weiss, W. W., Barban, C., et al. 2008b, ArXiv e-prints

Karoff, C. 2008, *PhD thesis*, Department of Physics and Astronomy, University of Aarhus, Denmark

Kjeldsen, H., & Bedding, T. R. 1995, A&A, 293, 87

Kjeldsen, H., Bedding, T. R., Arentoft, T., et al. 2008a, ApJ, 682, 1370

Kjeldsen, H., Bedding, T. R., & Christensen-Dalsgaard, J. 2008b, ApJ, 683, L175

Ludwig, H., Samadi, r., Steffen, M., et al. 2009, ArXiv e-prints

Mathur, S., Garcia, R. A., Regulo, C., et al. 2009, ArXiv e-prints

Nordlund, A., Spruit, H. C., Ludwig, H.-G., & Trampedach, R. 1997, A&A, 328, 229

Stello, D., Chaplin, W. J., Bruntt, H., et al. 2009, ArXiv e-prints

Stello, D., Kjeldsen, H., & Bedding, T. R. 2007, in Astronomical Society of the Pacific Conference Series, 366, Transiting Extrasolar Planets Workshop, ed. C. Afonso, D. Weldrake, & T. Henning, 247

Svensson, F., & Ludwig, H.-G. 2005, in ESA Special Publication, 560, 13th Cambridge Workshop on Cool Stars, Stellar Systems and the Sun, ed. F. Favata, G. A. J. Hussain, & B. Battrick, 979